YEARB

NUMBER TWENTY-TWO

Yearbook 2019–20

United Kingdom Mathematics Trust

United Kingdom Mathematics Trust

Yearbook 2019–20

Published by the United Kingdom Mathematics Trust.

School of Mathematics, University of Leeds,

Leeds, LS2 9JT, United Kingdom

www.ukmt.org.uk

First published 2020.

ISBN 978-1-906001-37-7

Printed in the UK for the UKMT by Charlesworth Press, Wakefield.

www.charlesworth.com

Typographic design by Andrew Jobbings.

Typeset with LaTeX.

The books published by the United Kingdom Mathematics Trust are grouped into series.

The EXCURSIONS IN MATHEMATICS series consists of monographs which focus on a particular topic of interest and investigate it in some detail, using a wide range of ideas and techniques. They are aimed at high school students, undergraduates and others who are prepared to pursue a subject in some depth, but do not require specialised knowledge.

1. *The Backbone of Pascal's Triangle*, Martin Griffiths
2. *A Prime Puzzle*, Martin Griffiths

The HANDBOOKS series is aimed particularly at students at secondary school who are interested in acquiring the knowledge and skills which are useful for tackling challenging problems, such as those posed in the competitions administered by the UKMT and similar organisations.

1. *Plane Euclidean Geometry: Theory and Problems*, A D Gardiner and C J Bradley
2. *Introduction to Inequalities*, C J Bradley
3. *A Mathematical Olympiad Primer*, Geoff C Smith
4. *Introduction to Number Theory*, C J Bradley
5. *A Problem Solver's Handbook*, Andrew Jobbings
6. *Introduction to Combinatorics*, Gerry Leversha and Dominic Rowland
7. *First Steps for Problem Solvers*, Mary Teresa Fyfe and Andrew Jobbings
8. *A Mathematical Olympiad Companion*, Geoff C Smith
9. *Topics in Combinatorics*, Gerry Leversha and Dominic Rowland

The PATHWAYS series aims to provide classroom teaching material for use in secondary schools. Each title develops a subject in more depth and in more detail than is normally required by public examinations or national curricula.

1. *Crossing the Bridge*, Gerry Leversha
2. *The Geometry of the Triangle*, Gerry Leversha

The PROBLEMS series consists of collections of high-quality and original problems of Olympiad standard.

1. *New Problems in Euclidean Geometry*, David Monk

The CHALLENGES series is aimed at students at secondary school who are interested in tackling stimulating problems, such as those posed in the Mathematical Challenges administered by the UKMT and similar organisations.

1. *Ten Years of Mathematical Challenges: 1997 to 2006,*
2. *Ten Further Years of Mathematical Challenges: 2006 to 2016,*
3. *Intermediate Problems,* Andrew Jobbings
4. *Junior Problems,* Andrew Jobbings
5. *Senior Problems,* Andrew Jobbings

The YEARBOOKS series documents all the UKMT activities, including details of all the challenge papers and solutions, accounts of the IMO and Olympiad training camps, and other information about the Trust's work during each year.

Contents

Foreword by the Chair

This is my first foreword to the UKMT Yearbook, as I took over as Chair from Professor Chris Budd in August 2019. I would like to start by thanking Chris for all the hard work he devoted to UKMT during his years as Chair. He left the organisation in excellent shape, both financially and from a governance point of view, and this has proved invaluable as we have navigated the challenges of the last twelve months.

Our first challenge in 2019/20 was to rebuild the executive team in Leeds following the departure of Rachel Greenhalgh, Director and Steven O'Hagan, Deputy Director. Following a formal recruitment process, we were delighted to appoint Hannah Telfer, previously Operations Manager, to the role of Director. Hannah moved swiftly, which with hindsight was wise, to strengthen her team and I would like to welcome to UKMT, Amanda Lassu and Chris Normington (Operations Managers) and Emma Pilkington (Volunteering Manager) Thank you, Hannah, Amanda, Chris and Emma, for taking on these challenging roles, and to Rachel and Steven for their contributions over the years.

The UKMT volunteers are central to the success of the organisation and Emma has many plans to enhance our communication and interaction with all our volunteers. We also have plans in place to give volunteers a greater say in the governance of UKMT. These have been inevitably delayed by Covid but we hope to implement them shortly.

It is, of course, impossible to write this without covering Covid. The national lockdown in March had a major impact on our activities. The office in Leeds was closed, and as I write remains closed as it is located within the University of Leeds campus. The team adjusted admirably to working from home and such activities as could take place have been managed successfully.

With schools being shut from March it was not possible to run the Junior Mathematical Challenge in the usual way, but we were able to switch rapidly to an online format. This was well received by many schools and took place successfully in June. Looking forward to 2020/21 we intend to offer schools the flexibility to choose between paper-based or online versions of the Challenges, in response to the feedback we received.

Sadly, it was impossible to continue with the Team Challenges as the format of these competitions do not allow for social distancing, as anyone who has attended one will testify! We will bring these competitions back once the national situation permits.

Preparation for, and participation in, international Maths competitions continued, albeit with a switch to virtual activities. We entered a team in the Romanian Master of Mathematics which was able to take place shortly before lockdown and in the virtual European Girls Maths Olympiad. The International Mathematics Olympiad 2020 itself took place in September, hosted virtually by Russia. The UK team achieved an excellent result coming ninth out of 105 participating countries and earning one gold, four silvers and one bronze medal, including the only gold won by a female competitor.

Although our outreach activities were also affected by Covid, we were able to run a 'virtual' Summer School in Summer 2020 and are putting plans in place to repeat this in 2021 if necessary. Mentoring has been able to continue.

As we are currently unable to despatch our publications from Leeds, we have switched to marketing them via Amazon, where they remain available. During the lockdown, we also made all our past papers available free via our website, www.UKMT.org.uk.

During the year, we were delighted to welcome XTX Markets as our new principal supporter. We thank them for their invaluable backing and also thank all our existing and past supporters including Oxford Asset Management, LetterOne, Man Group and the Institute and Faculty of Actuaries.

Finally, I would like to thank all the Trustees of UKMT for their help and support, and in particular to thank Dr Geoff Smith who stood down as Deputy Chair during the year after many years on the Board, although he remains President of the IMO Board and closely involved in the Olympiads.

Graham Keniston-Cooper MA, MMath (Cantab) `chair@ukmt.org.uk`

Introduction from the Executive Director

2020 has proved challenging, exhausting and thrown up many obstacles in both our working and home lives. Some of us have been directly impacted by the effects of Covid-19, others have experienced extreme pressures whilst also undergoing the trials and tribulations of "normal" life. We have missed interactions with our families, friends and work colleagues, and have had to fundamentally shift our whole way of existence to respond to the ever changing government guidance. Despite this, the UKMT has still achieved some amazing things and I would like to pause a moment to celebrate this. It is testament to the hard work and commitment of our schools, volunteers, staff and sponsors which has allowed the trust to continue fulfilling its charitable aim of advancing the education of young people in mathematics in these most turbulent times. A heartfelt thank you to each and everyone that has contributed over the last year, it really has made a difference.

This is my first introduction as Executive Director, and as I write this, I reflect upon the year feeling proud of all UKMT has accomplished. When the UKMT offices based at the University of Leeds were closed in late March 2020, I never dreamt we would still be working from home a year later. Remote working for colleagues and volunteers has necessitated an ability to adapt and to explore new ways of working. This has been a steep learning curve and often delivered at breakneck speed, but we have emerged as a stronger and more agile organisation as a result.

One of my particular highlights from the year was around the Junior Mathematical Challenge (JMC). As school closures threatened our ability to run the JMCe, staff and volunteers rapidly pulled together to move the event online and we witnessed students taking the JMC from home for the first time in UKMT's history.

In summary, 2019-20 has been a year of significant change due to the pandemic. It has been a pleasure to see UKMT continue to advance the education of young people in mathematics. I look forward to building on the lessons and successes of last year as we hopefully move towards a happier and more positive 2021/22.

Hannah Telfer
Executive Director
director@ukmt.org.uk

Chapter 1

History of UKMT

1.1 Foundation

National mathematics competitions have existed in the UK for several decades. Up until 1987 the total annual participation was around 8000. There was then an enormous growth, from 24,000 in 1988 to around 250,000 in 1995 – without doubt due to the drive, energy and leadership of Dr Tony Gardiner who, at that time, was a mathematician at the University of Birmingham. By the end of this period there were some nine or ten competitions for UK schools and their students organised by three different bodies: the British Mathematical Olympiad Committee, the National Committee for Mathematical Contests and the UK Mathematics Foundation.

During 1995, discussions took place between interested parties to seek a way of setting up a single body to continue and develop these competitions and related activities. This led to the formation of the United Kingdom Mathematics Trust (UKMT), which was incorporated as a company limited by guarantee (no. 03271283) in October 1996 and registered with the Charity Commission for England and Wales (no. 1059125).

Throughout its existence, the UKMT has continued to develop and expand the number of mathematics competitions and events. As a result, around 700,000 students throughout the UK, and beyond, now participate in the Mathematical Challenges alone, and their teachers (as well as others) not only provide much valued help and encouragement, but also take advantage of the support offered to them by the Trust.

1.2 Aims of the Trust

The charitable purpose of the UKMT is 'to advance the education of children and young people in mathematics'. To attain this, it is empowered to engage in activities ranging from teaching to publishing, research and lobbying. Its focal point is the organisation of mathematics competitions, from popular mass Mathematical Challenges and Team Maths Challenges events to the selection and training of the UK team for the annual International Mathematical Olympiad.

Chapter 2

Solo competitions

2.1 Overview

Our suite of competitions for individuals is designed to provide students, and their teachers, with a source of challenging and interesting mathematical problems. They aim to help develop students' appreciation and enjoyment of mathematical thinking and problem solving.

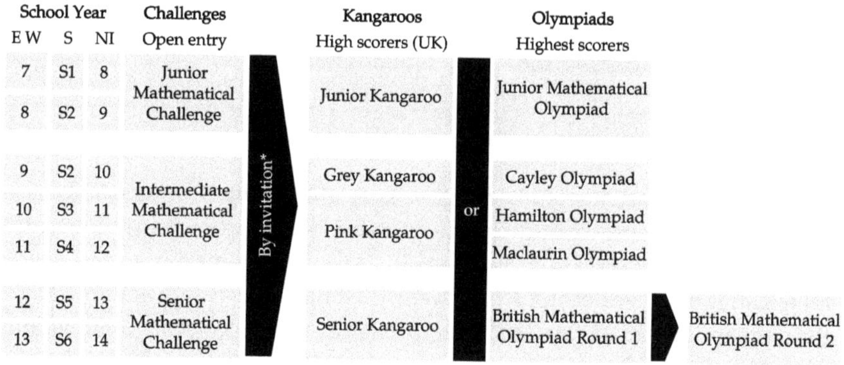

School Year EW	S	NI	Challenges Open entry	Kangaroos High scorers (UK)	Olympiads Highest scorers	
7	S1	8	Junior Mathematical Challenge	Junior Kangaroo	Junior Mathematical Olympiad	
8	S2	9				
9	S2	10	Intermediate Mathematical Challenge	Grey Kangaroo	Cayley Olympiad	
10	S3	11		Pink Kangaroo	Hamilton Olympiad	
11	S4	12			Maclaurin Olympiad	
12	S5	13	Senior Mathematical Challenge	Senior Kangaroo	British Mathematical Olympiad Round 1	British Mathematical Olympiad Round 2
13	S6	14				

By invitation*

or

All but one of our solo competitions (the Mathematical Olympiad for Girls) fit into a structure largely based on the school year students are in. The figure shows, for each secondary-school year in England, Wales, Scotland

* In case of extenuating circumstances, students may be entered into the Olympiads as Discretionary Candidates for a fee.

and Northern Ireland, the various competitions open to students at that stage in their education. In most cases, students may enter competitions aimed at more advanced school years at the discretion of their school.

We now give a brief overview of each of our solo competitions. The remainder of this chapter consists of competition materials from the academic year 2019-20.

2.1.1 Mathematical Challenges

We intend that problems on the Junior, Intermediate and Senior Mathematical Challenges should not require any mathematical knowledge beyond the four national curricula of the UK. Instead, the problems are designed to promote mathematical reasoning, precision of thought, and fluency in using basic mathematical techniques to solve interesting problems. Most problems are intended to be accessible, yet still challenge those with more experience; they are also meant to be memorable and enjoyable.

Each of these three competitions has 25 problems, each with five multiple-choice answers. The first 15 problems are intended to be accessible to all participants with the final 10 providing more of a challenge. Schools were given the option for students to take the Challenges in their schools or colleges and indicate their answers on a machine-readable answer sheet, which is returned to us for marking. Due to school closures in March 2019, JMC participants were given the option to participate by paper, or by completing the Challenge online.

2.1.2 Kangaroos

The UKMT is the UK member of the *Association Kangourou sans Frontières*, a collaborative association whose purpose is to promote mathematics among young people around the world. Around six million young people take Kangaroo competitions each year. The name *Kangaroo* acknowledges the contribution made by Australia to establishing large-scale mathematics contests.

We offer the Junior, Grey, Pink and Senior Kangaroo competitions as follow-on rounds for certain high scorers in the Mathematical Challenges who have not scored quite highly enough to qualify for the relevant Olympiad.

The Junior, Grey and Pink Kangaroos each have 25 problems, each with five multiple-choice answers. Each answer in the Senior Kangaroo is an integer from 0 to 999. Students sit Kangaroo papers in their schools or colleges and indicate their answers on a machine-readable answer sheet, which is returned to us for marking.

2.1.3 Olympiads

Both candidates and their teachers will find it helpful to know something of the general principles involved in marking Olympiad-type papers. The preliminary paragraphs therefore serve as an exposition of the 'philosophy' which has guided both the setting and marking of all such papers at all age levels, both nationally and internationally.

What we are looking for, essentially, are solutions to problems. This approach is therefore rather different from what happens in public examinations such as GCSE and A level, where credit is given for the ability to carry out individual techniques regardless of how these techniques fit into a protracted argument. Such marking is cumulative; a candidate may gain 60% of the available marks without necessarily having a clue about how to solve the final problem. Indeed, the questions are generally structured in such a way as to facilitate this approach, divided into many parts and not requiring an overall strategy for tackling a multi-stage argument.

In distinction to this, Olympiad-style problems are marked by looking at each question synoptically and deciding whether the candidate has some sort of overall strategy or not. An answer which is essentially a solution, but might contain either errors of calculation, flaws in logic, omission of cases or technical faults, will be marked on a '10 minus' basis. One question we often ask is: if we were to have the benefit of a two-minute interview with this candidate, could they correct the error or fill the gap? On the other hand, an answer which shows no sign of being a genuine solution is marked on a '0-plus' basis; up to 3 marks might be awarded for particular cases or insights. It is therefore important that candidates taking these papers realise the importance of the rubric about trying to finish whole questions rather than attempting lots of disconnected parts.

2.2 Junior Mathematical Challenge

2.2.1 Question paper and solutions

United Kingdom
Mathematics Trust

JUNIOR MATHEMATICAL CHALLENGE

Thursday 30 April 2020

supported by 

England & Wales: Year 8 or below
Scotland: S2 or below
Northern Ireland: Year 9 or below

INSTRUCTIONS

1. Do not open the paper until the invigilator tells you to do so.
2. Time allowed: **60 minutes**.
   No answers, or personal details, may be entered after the allowed time is over.
3. The use of blank or lined paper for rough working is allowed; **squared paper, calculators and measuring instruments are forbidden.**
4. **Use a B or an HB non-propelling pencil.** Mark at most one of the options A, B, C, D, E on the Answer Sheet for each question. Do not mark more than one option.
5. **Do not expect to finish the whole paper in the time allowed.** The questions in this paper have been arranged in approximate order of difficulty with the harder questions towards the end. You are not expected to complete all the questions during the time. You should bear this in mind when deciding which questions to tackle.
6. **Scoring rules:**
   5 marks are awarded for each correct answer to Questions 1-15;
   6 marks are awarded for each correct answer to Questions 16-25.
   In this paper you will not lose marks for getting questions wrong.
7. Your Answer Sheet will be read by a machine. **Do not write or doodle on the sheet except to mark your chosen options.** The machine will read all black pencil markings even if they are in the wrong places. If you mark the sheet in the wrong place, or leave bits of eraser stuck to the page, the machine will interpret the mark in its own way.
8. **The questions on this paper are designed to challenge you to think, not to guess.** You will gain more marks, and more satisfaction, by doing one question carefully than by guessing lots of answers. This paper is about solving interesting problems, not about lucky guessing.

Enquiries about the Junior Mathematical Challenge should be sent to:

UK Mathematics Trust, School of Mathematics, University of Leeds, Leeds LS2 9JT

☎ 0113 343 2339 enquiry@ukmt.org.uk www.ukmt.org.uk

1. Exactly one of the following five numbers is *not* prime. Which is it?

A 101 B 103 C 107 D 109 E 111

2. What is the value of $2020 \div 20$?

A 10 B 11 C 100 D 101 E 111

3. Each of these figures is based on a rectangle whose centre is shown.

How many of the figures have rotational symmetry of order two?

A 1 B 2 C 3 D 4 E 5

4. How many centimetres are there in 66.6 metres?

A 66600 B 6660 C 666 D 66.6 E 66

5. Amrita thinks of a number. She doubles it, adds 9, divides her answer by 3 and finally subtracts 1. She obtains the same number she originally thought of.
What was Amrita's number?

A 1 B 2 C 3 D 4 E 6

6. What is the value of $\frac{6}{12} - \frac{5}{12} + \frac{4}{12} - \frac{3}{12} + \frac{2}{12} - \frac{1}{12}$?

A $\frac{1}{2}$ B $\frac{1}{3}$ C $\frac{1}{4}$ D $\frac{1}{5}$ E $\frac{1}{6}$

7. Four different positive integers have a product of 110. What is the sum of the four integers?

A 19 B 22 C 24 D 25 E 28

8. Wesley has a grid of six cells. He wants to colour two of the cells black so that the two black cells share a vertex but not a side. In how many ways can he achieve this?

A 2 B 3 C 4 D 5 E 6

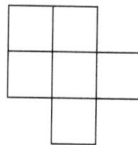

9. One half of one third of one quarter of one fifth of a number is 2.
What is the number?

A 240 B 120 C 60 D $\frac{1}{120}$ E $\frac{1}{240}$

10. How many of these equations have the solution $x = 12$?

$x - 2 = 10 \quad \frac{x}{2} = 24 \quad 10 - x = 2 \quad 2x - 1 = 25$

A 4 B 3 C 2 D 1 E 0

11. This 3 by 3 grid shows nine 1 cm × 1 cm squares and uses 24 cm of wire. What length of wire is required for a similar 20 by 20 grid?

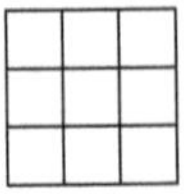

A 400 cm   B 420 cm   C 441 cm   D 800 cm   E 840 cm

12. The diagram shows an equilateral triangle divided into four smaller equilateral triangles. One of these triangles has itself been divided into four smaller equilateral triangles.
What fraction of the area of the large triangle has been shaded?

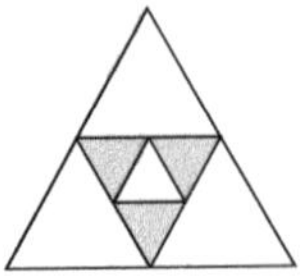

A $\frac{1}{8}$ B $\frac{3}{16}$ C $\frac{1}{4}$ D $\frac{5}{16}$ E $\frac{3}{8}$

13. The mean of four positive integers is 5. The median of the four integers is 6.
What is the mean of the largest and smallest of the integers?

A 3   B 4   C 5   D 6   E 8

14. In the diagram, angle OLM is twice as large as angle PON. What is the size of angle OLM?

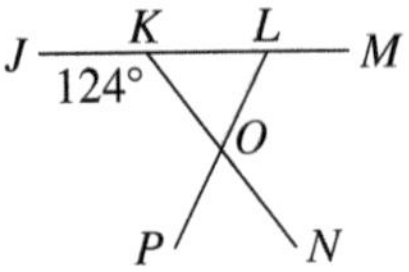

A 102°   B 106°   C 108°   D 112°   E 124°

J K L M
124°
O
P N

15. A group of 42 children all play tennis or football, or both sports. The same number play tennis as play just football. Twice as many play both tennis and football as play just tennis.
How many of the children play football?

A 7   B 14   C 21   D 28   E 35

16. You are given the sequence of digits "0625", and can insert a decimal point at the beginning, at the end, or at any of the other three positions.
Which of these numbers can you *not* make?

A $\frac{6}{25}$ B $\frac{5}{8}$ C $\frac{1}{16}$ D $\frac{25}{4}$ E 25^2

17. In 1925, Zbigniew Morón published a rectangle that could be dissected into nine different sized squares as shown in the diagram. The lengths of the sides of these squares are 1, 4, 7, 8, 9, 10, 14, 15 and 18.
What is the area of Morón's rectangle?

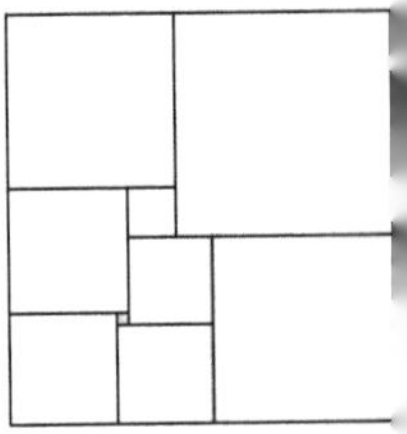

A 144   B 225   C 900   D 1024   E 1056

18. How many two-digit primes have both their digits non-prime?

A 6   B 5   C 4   D 3   E 2

19. In the table shown, the sum of each row is shown to the right of the row and the sum of each column is shown below the column.
What is the value of L?

A 1 B 2 C 3 D 5 E 7

J	K	J	5
K	K	L	13
L	J	L	15
11	7	15	

20. Edmund makes a cube using eight small cubes. Samuel uses cubes of the same size as the small cubes to make a cuboid twice as long, three times as wide and four times as high as Edmund's cube.
How many more cubes does Samuel use than Edmund?

A 9 B 24 C 64 D 184 E 190

21. The digits of both the two-digit numbers in the first calculation below have been reversed to give the two-digit numbers in the second calculation. The answers to the two calculations are the same.

$$62 \times 13 = 806 \qquad 26 \times 31 = 806$$

For which one of the calculations below is the same thing true?

A 25×36 B 34×42 C 54×56 D 42×48 E 32×43

22. Harriet has a square piece of paper. She folds it in half to form a rectangle and then in half again to form a second rectangle (which is not a square). The perimeter of the second rectangle is 30 cm.
What is the area of the original square?

A $36\,\text{cm}^2$ B $64\,\text{cm}^2$ C $81\,\text{cm}^2$ D $100\,\text{cm}^2$ E $144\,\text{cm}^2$

23. There is more than one integer, greater than 1, which leaves a remainder of 1 when divided by each of the four smallest primes.
What is the difference between the two smallest such integers?

A 211 B 210 C 31 D 30 E 7

24. Susan is attending a talk at her son's school. There are 8 rows of 10 chairs where 54 parents are sitting. Susan notices that every parent is either sitting on their own or next to just one other person.
What is the largest possible number of adjacent empty chairs in a single row at that talk?

A 3 B 4 C 5 D 7 E 8

25. In the diagram, $PQRS$, JQK and LRK are straight lines.

What is the size of the angle JKL?

A 34° B 35° C 36° D 37° E 38°

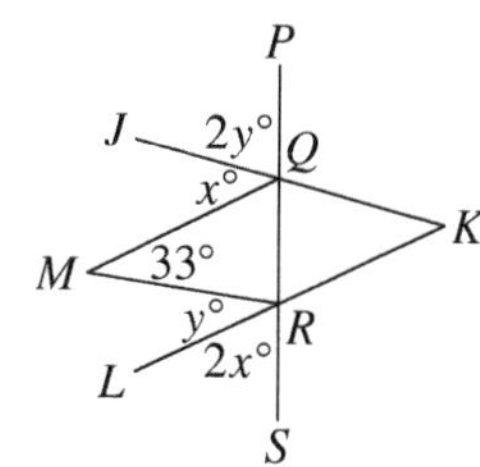

United Kingdom Mathematics Trust

Junior Mathematical Challenge

Thursday 30 April 2020

For reasons of space, these solutions are necessarily brief.

There are more in-depth, extended solutions available on the UKMT website, which include some exercises for further investigation:

www.ukmt.org.uk

1. E It is clear that 111 is a multiple of 3 since the sum of its digits is 3. Therefore 111 is not prime.

2. D The value of $2020 \div 20$ is equal to the value of $202 \div 2 = 101$.

3. C The only rotational symmetry of the rectangle is rotation through $180°$ and so has order two. The diagrams below show the effect of this rotation on each of the five figures.

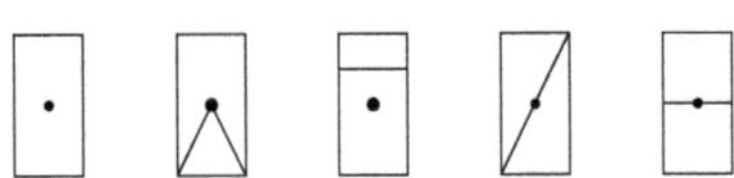

Only the first, fourth and fifth of these figures remain unchanged and so just three figures have rotational symmetry of order two.

4. B There are 100 cm in 1 m. So the number of centimetres in 66.6 m is $100 \times 66.6 = 6660$.

5. E Let Amrita's number be n. The information in the question tells us that $(2n + 9) \div 3 - 1 = n$. Therefore $2n + 9 = 3(n + 1) = 3n + 3$. Hence $n = 9 - 3 = 6$.

6. C The value of $\frac{6}{12} - \frac{5}{12} + \frac{4}{12} - \frac{3}{12} + \frac{2}{12} - \frac{1}{12} = \frac{1}{12} + \frac{1}{12} + \frac{1}{12} = \frac{3}{12} = \frac{1}{4}$.

7. A Note that $110 = 10 \times 11 = 2 \times 5 \times 11$, which is the prime factorisation of 110. Therefore the four different positive integers whose product is 110 are 1, 2, 5 and 11. Their sum is 19.

8. D The diagram shows the original diagram with the cells labelled 1, 2, 3, 4, 5, 6. The pairs of cells which Wesley can colour are 1 and 4; 2 and 3; 2 and 5; 3 and 6; 5 and 6. Therefore the diagram can be coloured in five ways.

1	2	
3	4	5
	6	

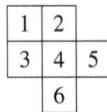

9. A Let the required number be n. Then, $\frac{1}{2} \times \frac{1}{3} \times \frac{1}{4} \times \frac{1}{5} \times n = 2$. Therefore $\frac{1}{120} \times n = 2$. Hence $n = 120 \times 2 = 240$.

10. D The solutions of the four equations are, from left to right, $x = 12$; $x = 48$; $x = 8$; $x = 13$. Therefore exactly one of the equations has solution $x = 12$.

11. E In a 20 by 20 grid there are 21 horizontal wires each of length 20 cm and 21 vertical wires each of length 20 cm.
Therefore the total length of wire required for a 20 by 20 grid is $42 \times 20\,\text{cm} = 840\,\text{cm}$.

12. B The large equilateral triangle has been divided into four congruent equilateral triangles. Therefore each of these triangles has an area equal to one quarter of the area of the large triangle. Similarly, the equilateral triangle which is partly shaded has been divided into four congruent equilateral triangles, three of which are shaded.
Therefore the fraction of the area of the large triangle which has been shaded is $\frac{3}{4} \times \frac{1}{4} = \frac{3}{16}$.

13. B The mean of four positive integers is 5. Therefore the sum of the four integers is $4 \times 5 = 20$.
The median of the integers is the mean of the two middle integers.
Since this median is 6, the sum of the two middle integers is $2 \times 6 = 12$.
Hence the sum of the smallest and largest of the four integers is $20 - 12 = 8$.
Therefore the mean of the largest and smallest of the integers is $8 \div 2 = 4$.

14. D Let the size, in degrees, of $\angle OLM$ be $2x$. Then, in degrees, the size of $\angle PON$ is x.
Therefore, since vertically opposite angles are equal, $\angle KOL = x°$.
The sum of the angles on a straight line is 180°, so $\angle LKO = 180° - 124° = 56°$.
An exterior angle of a triangle is equal to the sum of the two opposite, interior angles.
Therefore, in triangle KOL, $2x = x + 56$. So $x = 56$.
Hence $\angle OLM$ is $2 \times 56° = 112°$.

15. E Let the number of children who play only football be f, the number of children who play only tennis be t and the number of children who play both sports be b.
Since there are 42 children, $f + t + b = 42$.
Also, since the number of children who play tennis is equal to the number of children who play only football, $t + b = f$. Therefore $f + f = 42$. So $f = 21$ and $t + b = 21$.
Finally, twice as many play both tennis and football as play just tennis. Therefore $b = 2t$.
Substituting for b, gives $t + 2t = 21$. Hence $t = 7$.
Therefore the number of children who play football is $42 - t = 42 - 7 = 35$.

16. A Note that $\frac{6}{25} = \frac{24}{100} = 0.24$. Therefore $\frac{6}{25}$ cannot be made from the sequence "0625".
Of the other four options, $\frac{5}{8} = 0.625$; $\frac{1}{16} = .0625$; $\frac{25}{4} = 06.25$; $25^2 = 0625$.
Hence $\frac{6}{25}$ is the only option which cannot be made from the given sequence.

17. E Label the individual squares, except for the smallest, P, Q, R, S, T, U, V and W as shown. The side-length of the smallest square is 1. Since the next smallest squares are W and V, they have side-lengths 4 and 7 respectively. Therefore Q has side-length $7 + 1 = 8$, P has side-length $8 + 1 = 9$ and U has side-length $9 + 1 = 10$. Hence T has side-length $10 + 4 = 14$ and R has side-length $7 + 8 = 15$. Therefore the rectangle has side-lengths $9 + 8 + 15 = 32$ and $9 + 10 + 14 = 33$. Therefore the area of the rectangle is $32 \times 33 = 1056$.

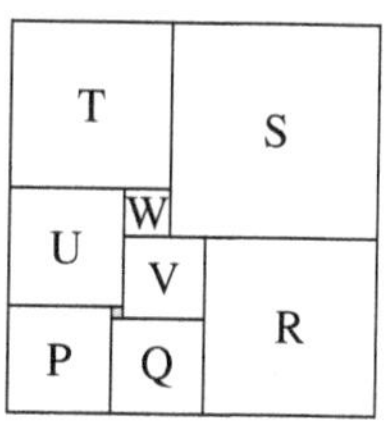

(Alternatively: the three largest squares R, S and T have side-length 15, 18 and 14 respectively. So the rectangle has sides of length 18 + 15 and 18 + 14 and so measures 33 by 32.)

18. B The digits which are non-prime are 0, 1, 4, 6, 8, 9. However, the units digit of a prime cannot be 0, 4, 6 or 8.
Therefore any two-digit primes which have both their digits non-prime have a units digit of 1 or 9.
The only such primes are 11, 19, 41, 61 and 89. Hence there five such primes.

19. E Adding the top row and the middle column gives,
$2J + K + 2K + J = 5 + 7 = 12$. Hence $3J + 3K = 12$. So $J + K = 4$.
The first column shows that $J + K + L = 11$.
Hence, $J + K + L - (J + K) = 11 - 4 = 7$. Therefore $L = 7$.
(It is then possible to deduce that $J = 1$ and $K = 3$ and check that each total is correct.)

J	K	J	5
K	K	L	13
L	J	L	15
11	7	15	

20. D Edmund makes a cube using eight small cubes. Therefore his cube measures $2 \times 2 \times 2$.
Since Samuel makes a cuboid twice as long, three times as wide and four times as high as Edmund's cube, Samuel's cube measures $4 \times 6 \times 8$. Therefore Samuel uses 192 small cubes in making his cube.
Hence Samuel uses $192 - 8 = 184$ more small cubes than Edmund.

21. D It is possible to eliminate certain options by comparing the units digit of the original product with the units digit of the new product.
The units digit of 25×36 is 0, but the units digit of 52×63 is 6. So A is not the correct option.
The corresponding units digits for the other options are B: 8 and 2; C: 4 and 5; D: 6 and 6; E: 6 and 2. These calculations suggest that the correct option is D.
In confirmation, note that $42 \times 48 = 2 \times 21 \times 2 \times 24 = 24 \times 2 \times 2 \times 21 = 24 \times 84$.

22. E Let the side-length of Harriet's square be $4x$ cm. Then, the rectangle obtained when the square is folded in half measures $4x$ cm by $2x$ cm.
When that rectangle is folded in half, it gives a rectangle which is not a square so this rectangle measures $4x$ cm by x cm. The perimeter of the smaller rectangle is $2(4x + x)\,\text{cm} = 10x\,\text{cm}$.
Therefore, since the perimeter of the smaller rectangle is given as 30 cm, we have $x = 3$.
Hence the side-length of Harriet's square is 12 cm and its area is $144\,\text{cm}^2$.

23. B The four smallest primes are 2, 3, 5, 7. Since these are all prime, their lowest common multiple is $2 \times 3 \times 5 \times 7 = 210$.
Therefore the smallest positive integer which leaves remainder 1 when divided by each of the four smallest primes is $210 + 1 = 211$. The next positive such integer is $2 \times 210 + 1 = 421$.
Therefore the required difference is $421 - 211 = 210$.

24. B For the number of adjacent empty chairs in a single row to be a maximum, all rows apart from that one must hold as many parents as possible.

P	P	*	P	P	*	P	P	*	P
P	P	*	P	P	*	*	*	*	P

In view of Susan's observation, the maximum number of parents sitting in any row is seven, as the upper diagram shows. Therefore, the maximum number of parents in seven of the eight rows is $7 \times 7 = 49$. The number in the remaining row is then $54 - 49 = 5$. In that case, and again in view of Susan's observation, the maximum number of adjacent empty chairs would be four as shown in the lower diagram.

25. **E** Let $a°$, $b°$, $c°$, $d°$ and $e°$ be the sizes of the angles shown in the diagram. Since angles on a straight line add to 180°, the points Q and R on line PS give the equations
$x + 2y + a = 180$ and $y + 2x + b = 180$.
Adding these two equations gives $3x + 3y + a + b = 360$.
The sum of the interior angles of a triangle is 180°.
Hence, in triangle QMR, $a + b + 33 = 180$.
Subtracting this equation from the previous one gives
$3x + 3y - 33 = 180$ and hence $x + y = 71$.
Since vertically opposite angles are equal, we have $d = 2y$ and $e = 2x$.
Also, from triangle QRK we see that $d + e + c = 180$.
Therefore $2x + 2y + c = 180$ and hence $142 + c = 180$.
So the size of angle JKL is 38°.

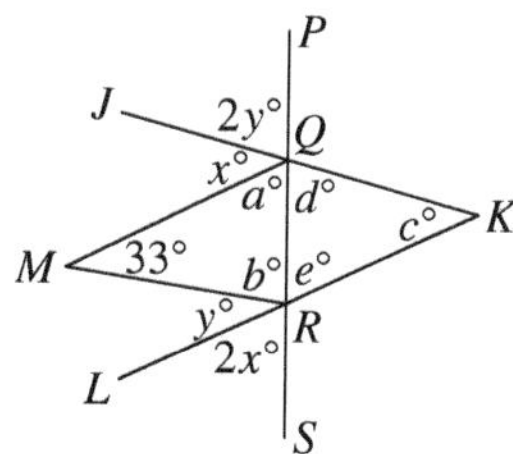

2.2.2 Participation

Participation in the Junior Mathematical Challenge

Year of competition	2016	2017	2018	2019	2020
Centres placing orders	4,002	4,064	4,115	4,152	3,113
Centres returning scripts	3,870	3,879	3,983	3,947	2,306
Papers ordered	303,020	307,280	317,260	322,120	253, 020
Scripts returned	263,371	265,884	276,638	273,984	129,211

Percentage change from previous year

Year of competition	2017	2018	2019	2020
Centres placing orders	+1.5%	+1.3%	+0.9%	-25.02%
Centres returning scripts	+0.2%	+2.7%	-0.9%	-41.57%
Papers ordered	+1.4%	+3.2%	+1.5%	-21.45%
Scripts returned	+1.0%	+4.0%	-1.0%	-52.84%

2.2.3 Student performance

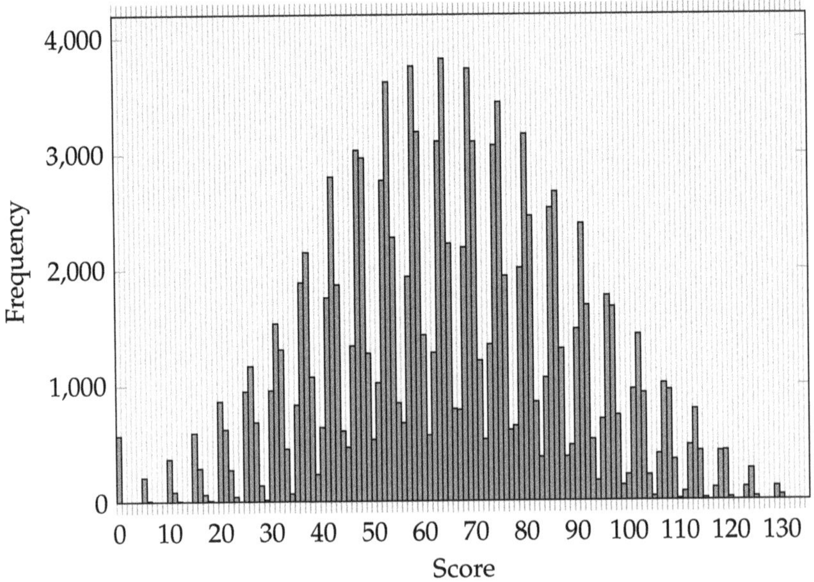

Answers given

The table below shows, for each question, the percentage of students giving each answer. The correct answer is underlined.

Question	A	B	C	D	E	Blank	Ambiguous
1	9	5	6	8	72	2	0
2	0	6	1	92	1	1	0
3	6	39	38	9	5	2	0
4	8	84	7	0	0	1	0
5	1	3	22	5	68	1	0
6	1	5	87	1	4	2	0
7	36	18	15	10	12	9	0
8	9	24	15	43	6	2	0
9	67	13	8	6	3	3	0
10	1	7	29	61	1	1	0
11	13	28	10	15	27	7	0
12	4	64	11	3	14	3	0
13	5	45	21	15	5	9	0
14	3	7	12	41	28	9	0
15	14	28	25	20	7	7	0
16	46	8	12	9	17	8	0
17	3	7	8	16	57	9	0
18	13	31	19	15	14	8	0
19	1	2	6	8	76	6	0
20	3	17	25	40	5	9	0
21	10	8	6	55	8	12	0
22	5	16	12	17	35	15	0
23	6	21	22	13	14	23	0
24	21	20	20	12	11	17	0
25	9	16	23	15	11	25	0

Mean score, award boundaries and follow-on thresholds

To recognise the highest performers, we award the top-scoring 40 % of participants Bronze, Silver and Gold certificates in the ratio 3 : 2 : 1, and we invite around 1200 of the very highest performers to take part in the Junior Mathematical Olympiad and around 7000 to take part in the Junior Kangaroo.

The table below shows the minimum score needed to obtain the corresponding award or follow-on round qualification for each of the previous five years of the competition.

Year of competition	2016	2017	2018	2019	2020
Junior Mathematical Olympiad Qual.	113	107	105	106	-
Junior Kangaroo Qual.	93	86	82	78	-
Gold	81	77	75	71	102
Silver	65	63	61	54	86
Bronze	51	52	49	41	70
Mean score	49.2	48.9	44.4	38	65.2

Comments from the Problems Group

Certificate thresholds, as many have noted, were rather higher than usual. A number of factors contributed to this:

- After a low average mark last year, the Problems Group aimed to set a paper that would be more widely accessible, especially in the earlier questions.
- Penalty marking was removed as an experiment because of a concern that it deters able candidates from attempting later questions.
- The conditions under which the JMC was sat were less rigorous than usual and we reflected this by designing a one-off certificate which clearly states 'online JMC' to differentiate it from other events.

2.3 Junior Kangaroo

Unfortunately, the Junior Kangaroo had to be cancelled this year due to coronavirus.

2.4 Junior Mathematical Olympiad

Unfortunately, the Junior Olympiad had to be cancelled this year due to coronavirus.

2.5 Intermediate Mathematical Challenge

2.5.1 Question paper and solutions

United Kingdom
Mathematics Trust

INTERMEDIATE MATHEMATICAL CHALLENGE

Thursday 6 February 2020

England & Wales: Year 11 or below
Scotland: S4 or below
Northern Ireland: Year 12 or below

INSTRUCTIONS

1. Do not open the paper until the invigilator tells you to do so.
2. Time allowed: **60 minutes**.
   No answers, or personal details, may be entered after the allowed time is over.
3. The use of blank or lined paper for rough working is allowed; **squared paper, calculators and measuring instruments are forbidden.**
4. **Use a B or an HB non-propelling pencil.** Mark at most one of the options A, B, C, D, E on the Answer Sheet for each question. Do not mark more than one option.
5. **Do not expect to finish the whole paper in the time allowed.** The questions in this paper have been arranged in approximate order of difficulty with the harder questions towards the end. You are not expected to complete all the questions during the time. You should bear this in mind when deciding which questions to tackle.
6. **Scoring rules:**
   5 marks are awarded for each correct answer to Questions 1-15;
   6 marks are awarded for each correct answer to Questions 16-25;
   Each incorrect answer to Questions 16-20 loses 1 mark;
   Each incorrect answer to Questions 21-25 loses 2 marks.
7. Your Answer Sheet will be read by a machine. **Do not write or doodle on the sheet except to mark your chosen options.** The machine will read all black pencil markings even if they are in the wrong places. If you mark the sheet in the wrong place, or leave bits of eraser stuck to the page, the machine will interpret the mark in its own way.
8. **The questions on this paper are designed to challenge you to think, not to guess.** You will gain more marks, and more satisfaction, by doing one question carefully than by guessing lots of answers. This paper is about solving interesting problems, not about lucky guessing.

Enquiries about the Intermediate Mathematical Challenge should be sent to:

UK Mathematics Trust, School of Mathematics, University of Leeds, Leeds LS2 9JT

☎ 0113 343 2339 enquiry@ukmt.org.uk www.ukmt.org.uk

1. What is the value of $2 - (3 - 4) - (5 - 6 - 7)$?

A 11 B 9 C 5 D −5 E −7

2. Which one of these is a multiple of 24?

A 200 B 300 C 400 D 500 E 600

3. What is the difference between 25% of £37 and 25% of £17?

A £4.25 B £5 C £6 D £7.50 E £9.25

4. What fraction of this diagram is shaded?

A $\frac{13}{32}$ B $\frac{1}{2}$ C $\frac{9}{16}$ D $\frac{5}{8}$ E $\frac{13}{16}$

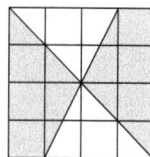

5. Four of the following coordinate pairs are the corners of a square.
Which is the odd one out?

A (4, 1) B (2, 4) C (5, 6) D (3, 5) E (7, 3)

6. Which of the following has the largest value?

A 2^6 B 3^5 C 4^4 D 5^3 E 6^2

7. Kartik wants to shade three of the squares in this grid blue and Lucy wants to shade the remaining two squares red. There are ten possible finished grids.
In how many of the finished grids are Lucy's red squares next to each other?

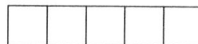

A 3 B 4 C 5 D 6 E 8

8. One of these options gives the value of $17^2 + 19^2 + 23^2 + 29^2$. Which is it?

A 2004 B 2008 C 2012 D 2016 E 2020

9. Adam's house number is in exactly one of the following ranges. Which one?

A 123 to 213 B 132 to 231 C 123 to 312 D 231 to 312 E 312 to 321

10. What is the value of $\frac{2468 \times 2468}{2468 + 2468}$?

A 2 B 1234 C 2468 D 4936 E 6 091 024

11. I start at square "1", and have to finish at square "7", moving at each step to a higher numbered adjacent square.
How many possible routes are there?

	2	4	6
1	3	5	7

A 7 B 9 C 10 D 11 E 13

12. Farmer Fatima rears chickens and goats. Today she returned from market and said, "I sold 80 animals, and now there are 200 fewer legs on my farm than before!"
How many goats did she sell?

A 15 B 20 C 25 D 30 E 35

13. What is half of 1.6×10^6?

A 8×5^6 B 4×10^6 C 8×10^5 D 8×10^2 E 1.6×10^3

14. The result of the calculation $9 \times 11 \times 13 \times 15 \times 17$ is the six-digit number '$3n8185$'.
What is the value of n?

A 2 B 4 C 6 D 8 E 0

15. Triangle PQR has been divided into twenty-five congruent right-angled triangles, as shown. The length of RP is 2.4 cm.

What is the length of PQ?

A 3 cm B 3.2 cm C 3.6 cm D 4 cm
E 4.8 cm

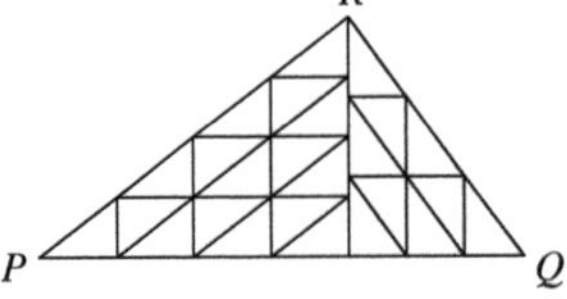

16. As a decimal, what is the value of $\frac{1}{9} + \frac{1}{11}$?

A 0.10 B 0.20 C 0.2020 D 0.202 020 E $0.\dot{2}\dot{0}$

17. The Knave of Hearts stole some tarts. He ate half of them, and half a tart more. The Knave of Diamonds ate half of what was left, and half a tart more. Then the Knave of Clubs ate half of what remained, and half a tart more. This left just one tart for the Knave of Spades.
How many tarts did the Knave of Hearts steal?

A 63 B 31 C 19 D 17 E 15

18. The diagram shows an isosceles right-angled triangle which has a hypotenuse of length y. The interior of the triangle is split up into identical squares and congruent isosceles right-angled triangles.
What is the total shaded area inside the triangle?

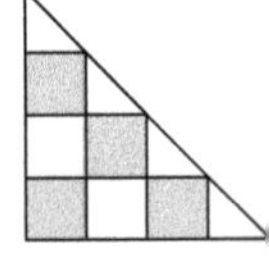

A $\frac{y^2}{2}$ B $\frac{y^2}{4}$ C $\frac{y^2}{8}$ D $\frac{y^2}{16}$ E $\frac{y^2}{32}$

19. The diagram shows two squares and four equal semicircles. The edges of the outer square have length 48 and the inner square joins the midpoints of the edges of the outer square. Each semicircle touches two edges of the outer square, and the diameter of each semicircle lies along an edge of the inner square.

What is the radius of each semicircle?

A 10 B 12 C 14 D 16 E 18

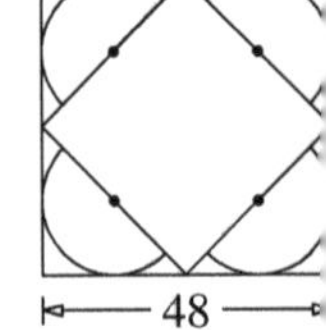

20. For any fixed value of x, which of the following four expressions has the largest value?

$(x+1)(x-1)$ $\quad$ $(x+\frac{1}{2})(x-\frac{1}{2})$ $\quad$ $(x+\frac{1}{3})(x-\frac{1}{3})$ $\quad$ $(x+\frac{1}{4})(x-\frac{1}{4})$

A $(x+1)(x-1)$ B $(x+\frac{1}{2})(x-\frac{1}{2})$ C $(x+\frac{1}{3})(x-\frac{1}{3})$ D $(x+\frac{1}{4})(x-\frac{1}{4})$
E it depends on the value of x

21. The diagram shows four semicircles, one with radius 2 cm, touching the other three, which have radius 1 cm.

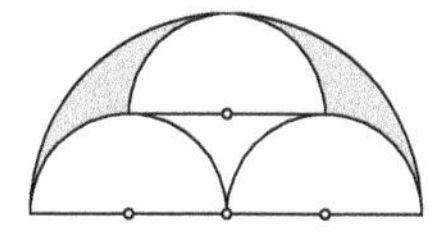

What is the total area, in cm^2, of the shaded regions?

A 1 B $\pi - 2$ C $2\pi - 5$ D $\frac{3}{2}$ E $\frac{1}{2}\pi$

22. The diagram shows a regular pentagon and an irregular quadrilateral.

What is the sum of the three marked angles?

A 72° B 90° C 108° D 126° E 144°

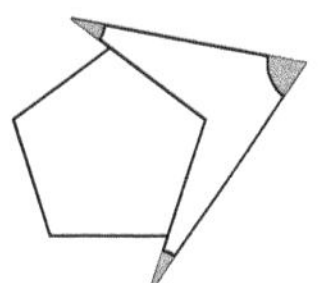

23. Five congruent triangles, each of which is half a square, are placed together edge to edge in three different ways as shown to form shapes P, Q and R.

P

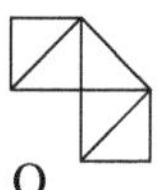
Q

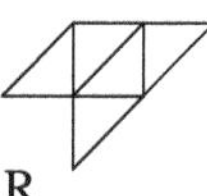
R

Which of the following lists gives the shapes in ascending order of the lengths of their perimeters?

A P, Q, R B Q, P, R C R, Q, P D R, P, Q E P, R, Q

24. The positive integers m and n are such that $10 \times 2^m = 2^n + 2^{n+2}$.
What is the difference between m and n?

A 1 B 2 C 3 D 4 E 5

25. The diagram shows six points P, Q, R, S, T and U equally spaced around a circle of radius 2 cm. The inner circle has radius 1 cm. The shaded region has three lines of symmetry.
Which of the following gives the area, in cm^2, of the shaded region?

A $2\pi + 3$ B $3\pi + 2$ C $\dfrac{4\pi + 3}{2}$ D $3(\pi + 2)$ E $4\pi + 3$

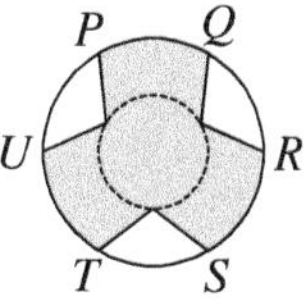

United Kingdom
Mathematics Trust

INTERMEDIATE MATHEMATICAL CHALLENGE

Solutions 2020

For reasons of space, these solutions are necessarily brief.

There are more in-depth, extended solutions available on the UKMT website, which include some exercises for further investigation:

www.ukmt.org.uk

1. A The value of $2-(3-4)-(5-6-7)=2-(-1)-(-1-7)=2+1-(-8)=2+1+8=11$.

2. E The correct answer is a multiple of both 3 and 8, as $24=3\times 8$. Although 100 is a multiple of 4, it is not a multiple of 8. Therefore even multiples of 100 are multiples of 8, but odd multiples are not. This leaves 200, 400 and 600, but of these only 600 is also a multiple of 3.
Hence the only option which is a multiple of 24 is 600.

3. B The required difference is 25% of (£37 - £17) = 25% of £20.
Hence the difference between 25% of £37 and 25% of £17 is £5.

4. D We choose units so the outer square has side-length 4, and hence area 16.
The unshaded area of the square consists of two congruent triangles of base 3 and height 2.
So the area of each unshaded triangle is $\frac{1}{2}\times 3\times 2=3$.
Hence the area of the shaded part of the square is $16-2\times 3=10$.
Therefore the fraction of the square which is shaded is $\frac{10}{16}=\frac{5}{8}$.

5. D First note that $AB=\sqrt{13}$, $AC=\sqrt{26}$, $AD=\sqrt{17}$ and $AE=\sqrt{13}$. Thus $AB=AE$.
Note also that $BC=\sqrt{13}$ and $CE=\sqrt{13}$. Therefore $ABCE$ is a rhombus.
Finally, $BE=\sqrt{26}$, so $ABCE$ is a rhombus with two equal diagonals and hence is a square.
So (3,5) is the odd point out.

6. C The values of the five options are 64, 243, 256, 125, 36 respectively.
Of these 256 is the largest, so 4^4 is the largest of the options.

7. B We label the five squares, from left to right, P, Q, R, S, T respectively.
In order to paint two adjacent squares, Lucy could paint P and Q, or Q and R, or R and S, or S and T. So in four of the finished grids, Lucy's red squares are adjacent to each other.

8. E Since the units digits of all the options are different, it is sufficient to add the units digits of the four squares in the sum. As $7^2=49$, 17^2 has a units digit of 9. In a similar way, we calculate that 19^2 has a units digit of 1, 23^2 has a units digit of 9 and 29^2 has a units digit of 1.
Therefore the units digit of the given sum is the units digit of $9+1+9+1$, that is 0.
As we are told that the correct answer is one of the options, we may deduc that it is 2020.

9. E Note that the numbers 123 to 213 are in both option A and option C. Also, numbers 213 to

231 are in both option B and option C, and numbers 231 to 312 are in both C and D. However, numbers 313 to 321 are in option E only. So that must be where Adam's house is.

10. B The value of $\frac{2468 \times 2468}{2468 + 2468}$ is $\frac{2468 \times 2468}{2 \times 2468} = \frac{2468}{2}$.

Hence the correct answer is 1234.

11. E There is exactly one route from "1" to "2", but two routes from "1" to "3" - one route directly from "1" to "3" and one route which moves from "1" to "2" to "3".
To move from "1" to "4", it is necessary to move from "2" directly to "4" or to move from "3" to "4". So the number of routes from "1" to "4" is $1 + 2 = 3$.
By a similar method, the number of routes from "1" to "5" is $2 + 3 = 5$ and from "1" to "6" is $3 + 5 = 8$. Finally, the number of routes from "1" to "7" is $5 + 8 = 13$.
(*Note that the numbers of routes to each of the squares is a term in the Fibonacci sequence.*)

12. B Let the number of chickens and goats sold be c and g respectively.
Then, since 80 animals are sold, we have $c + g = 80$.
Also, $2c + 4g = 200$, since the chickens have two legs and the goats have four legs.
Dividing the second equation by 2, we obtain $c + 2g = 100$.
Then, subtracting the first equation gives $g = 100 - 80 = 20$.
So 20 goats were sold.

13. C Half of 1.6×10^6 equals $0.8 \times 10^6 = 0.8 \times 10 \times 10^5 = 8 \times 10^5$.

14. A Note that $9 \times 11 \times 13 \times 15 \times 17$ is a multiple of 9. Therefore the sum of its digits is also a multiple of 9. The sum of the digits of '3n8185' equals $25 + n$. So $n = 2$, since 27 is a multiple of 9.

15. A The diagrams show one of the twenty-five congruent right-angled triangles and the two such triangles at vertex R.
Let the two acute angles in each of the triangles be $x°$ and $y°$.

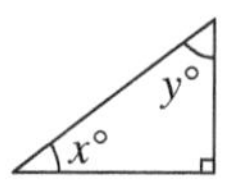

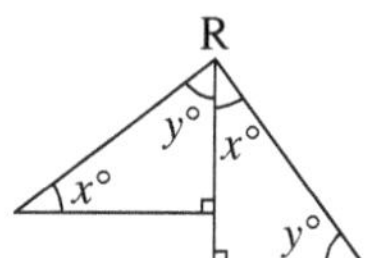

Note that $x + y + 90 = 180$, as the interior angles of a triangle sum to 180°.
So $x + y = 90$. In the second diagram, the two angles which meet at R are $x°$ and $y°$, so we can deduce that angle QRP is a right angle.
Let the length of the hypotenuse of each small triangle be a cm.
Note that angle QRP is a right angle, PR has length $4a$ cm and RQ has length $3a$ cm.
So the lengths of the sides in triangle PQR are in the ratio 3 : 4 : 5.
Therefore the length of PQ, in cm, is $5 \times \frac{2.4}{4} = 3$.

16. E The sum of $\frac{1}{9} + \frac{1}{11}$ is $\frac{11}{99} + \frac{9}{99}$, that is $\frac{20}{99}$.

Now 20 and 99 do not have any factors in common except 1. So the fraction cannot be simplified.
In option A, we may write 0.10 as the fraction $\frac{10}{100}$, whose denominator is a power of 10. The same is true of options B, C and D. Therefore, when simplified, none of these fractions can have 99 as a denominator.
It is left to the reader to confirm, by division, that $\frac{20}{99} = 0.\dot{2}\dot{0}$.
An alternative argument follows.
Let $x = 0.\dot{2}\dot{0} = 0.202020....$ Then $100x = 20.202020...$.
Subtracting the first equation from the second gives $99x = 20$.

So $0.\dot{2}\dot{0} = \frac{20}{99}$.

17. E Suppose that at a particular stage there are m tarts available for a Knave to eat and that there are n left after he has finished eating.
Then $n = m - (\frac{1}{2}m + \frac{1}{2}) = \frac{1}{2}m - \frac{1}{2}$.
Therefore, $m = 2n + 1$.
As the Knave of Spades received one tart, then the number of tarts which the Knave of Clubs was given was $2 \times 1 + 1 = 3$.
Similarly, the number of tarts which the Knave of Diamonds was given was $2 \times 3 + 1 = 7$.
Finally, the number of tarts which the Knave of Hearts stole was $2 \times 7 + 1 = 15$.

18. C Let the length of each equal side of the given triangle be x.
Then, by Pythagoras' theorem, $x^2 + x^2 = y^2$. So $x^2 = \frac{y^2}{2}$.
In the triangle, four of the squares are shaded while the unshaded area consists of two squares and four half-squares.
Therefore, half of the area of the triangle is shaded.
Now the area of the triangle is $\frac{1}{2} \times x \times x = \frac{1}{2}x^2$.
Therefore the total shaded area of the triangle is $\frac{1}{4}x^2 = \frac{1}{4} \times \frac{y^2}{2} = \frac{y^2}{8}$.

19. B The diagram shows the top left-hand corner of the original diagram. The centre of the semicircle shown is Q. Also, U and S are the points where the edges of the bigger square touch the semicircle shown. Therefore both QU and QS are radii of the semicircle and $\angle TSQ = \angle TUQ = 90°$. Also $\angle UTS$ is a right angle as it is the corner of a square. Therefore $UQST$ is a square. Hence $QS = ST$.

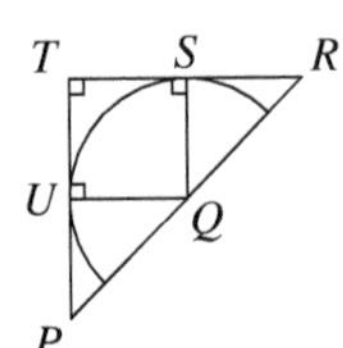

Note that $TR = TP$ because P and R are midpoints of the original large square. Therefore $\angle PRT = 45°$. So $QS = RS$.
Hence QS is half the length of TR, which is itself half of the length of a side of the outer square, which is 48 cm.
So the radius of the semicircle is one quarter of 48 cm = 12 cm.

20. D The first four expressions expand to $x^2 - 1$; $x^2 - \frac{1}{4}$; $x^2 - \frac{1}{9}$; $x^2 - \frac{1}{16}$ respectively.
Note that $\frac{1}{16} < \frac{1}{9} < \frac{1}{4} < 1$. Therefore the least value is $x^2 - \frac{1}{16}$, that is $(x + \frac{1}{4})(x - \frac{1}{4})$.
This result is irrespective of the value of x.

21. B In the diagram, SR is the diameter of the upper small semicircle and P and Q are the centres of the two lower small semicircles. Note that the line SR touches the two semicircles with centres P and Q at points S and R respectively.
So $\angle SRQ = \angle RSP = 90°$.
Also, $SR = PQ = 2$ cm. Therefore $PQRS$ is a rectangle.

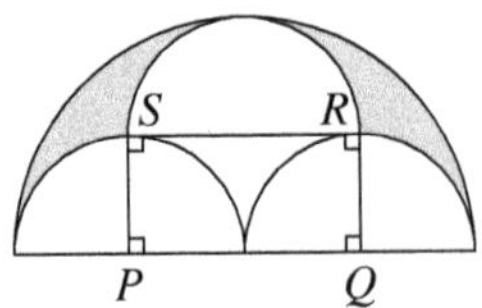

The total unshaded area in the diagram is the rectangle plus a semicircle and two quarter circles, that is, the rectangle plus a circle. So, in cm^2, it is $1 \times 2 + \pi \times 1^2 = 2 + \pi$.
So the total shaded area, in cm^2, is $\frac{1}{2} \times \pi \times 2^2 - (2 + \pi) = 2\pi - (\pi + 2) = \pi - 2$.
Hence the required area, in cm^2, is $\pi - 2$.

22. **C** The exterior angle of a regular pentagon is $\dfrac{360°}{5} = 72°$.
Therefore the interior angle of a regular pentagon, in degrees, is $180 - 72 = 108$. The angles at a point sum to 360°, so the reflex angle in the irregular quadrilateral, in degrees, is $360 - 108 = 252$. Finally the interior angles of a quadrilateral sum to 360°, so the sum of the marked angles, in degrees, is $360 - 252 = 108$.
(Note that the sum of the three marked angles equals the interior angle of the pentagon.)

23. **A** Let the small sides of each triangle have length r. This is also the side of the original square. The longer side of each triangle is $\sqrt{2}r$, since that is the diagonal of the square.
Hence the perimeters of shapes P, Q, R are $(2 + 3\sqrt{2})r$, $(6 + \sqrt{2})r$ and $(4 + 3\sqrt{2})r$ respectively.
Now $6 + \sqrt{2} - (2 + 3\sqrt{2}) = 4 - 2\sqrt{2} = 2(2 - \sqrt{2})$, which is greater than zero.
Therefore the perimeter of P is less than the perimeter of Q.
Also, $4 + 3\sqrt{2} - (6 + \sqrt{2}) = 2\sqrt{2} - 2 = 2(\sqrt{2} - 1)$, which is also greater than 0.
Therefore the perimeter of Q is less than the perimeter of R.
Hence, in ascending order, the lengths of the perimeters are P, Q, R.

24. **A** We are given that $10 \times 2^m = 2^n + 2^{n+2}$. Therefore $5 \times 2 \times 2^m = 2^n(1 + 2^2)$.
Hence $5 \times 2^{m+1} = 2^n \times 5$. So $m + 1 = n$. Therefore the difference between m and n is 1.

25. **A** The diagram shows exactly one third of the shaded area in the original diagram. It is made up of the quadrilateral $UOPX$, together with a sector of the outer circle, POQ, where O is the centre of the original circle.
Since P and Q are two of the six points equally spaced around the circle, $\angle POQ = \frac{1}{6} \times 360° = 60°$. The outer circle has radius 2 cm, so the area, in cm^2, of sector POQ is $\frac{1}{6} \times \pi \times 2^2 = \frac{2\pi}{3}$.

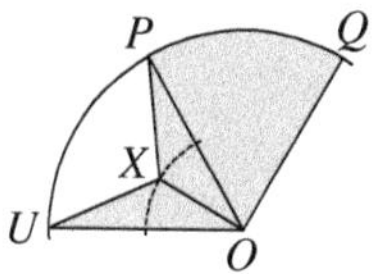

Since U and P are also equally spaced, $\angle UOP = 60°$. So $\angle UOX = 30°$.
Hence the area of triangle UOX, in cm^2, is $\frac{1}{2} \times 1 \times 2 \times \sin 30° = \frac{1}{2}$.
So the area of quadrilateral $UOPX$ is 1 cm^2.
Therefore the total shaded area, in cm^2, in the original diagram is $3 \times (\frac{2\pi}{3} + 1) = 2\pi + 3$.

2.5.2 Participation

Participation in the Intermediate Mathematical Challenge

Year of competition	2016	2017	2018	2019	2020
Centres placing orders	3,241	3,322	3,351	3,284	3,127
Centres returning scripts	3,140	3,225	3,250	3,159	3,002
Papers ordered	262,100	267,350	273,470	270,010	265,700
Scripts returned	215,258	219,969	226,356	222,292	217,111

Percentage change from previous year

Year of competition	2017	2018	2019	2020
Centres placing orders	+2.5%	+0.9%	-2.0%	-4.7%
Centres returning scripts	+2.7%	+0.8%	-2.8%	-4.9%
Papers ordered	+2.0%	+2.3%	-1.3%	-1.6%
Scripts returned	+2.2%	+2.9%	-1.8%	-2.33%

2.5.3 Student performance

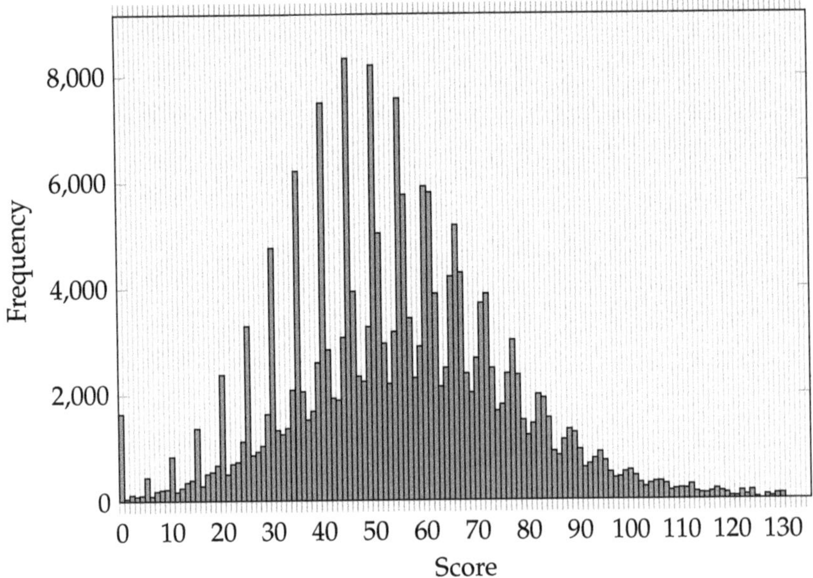

Answers given

The table below shows, for each question, the percentage of students giving each answer. The correct answer is underlined.

Question	A	B	C	D	E	Blank	Ambiguous
1	75	5	3	9	7	1	0
2	2	3	5	5	82	2	0
3	5	86	3	2	2	2	0
4	1	2	12	80	3	2	0
5	6	14	9	56	9	5	0
6	2	15	78	3	0	1	0
7	9	76	6	4	3	2	0
8	6	3	5	6	76	4	0
9	21	10	24	7	30	8	0
10	7	57	12	14	4	5	0
11	27	24	14	17	13	5	0
12	4	65	10	11	5	6	0
13	6	2	74	5	9	4	0
14	54	12	10	9	7	7	0
15	19	21	25	12	11	13	0
16	4	5	5	3	48	35	0
17	2	5	5	5	37	46	0
18	8	10	15	5	2	60	0
19	4	17	3	5	3	69	0
20	8	3	2	19	12	56	0
21	2	8	6	2	6	76	0
22	5	3	26	3	3	59	0
23	13	5	10	4	6	63	0
24	11	4	3	4	2	77	0
25	3	3	4	3	2	85	0

Mean score, award boundaries and follow-on thresholds

To recognise the highest performers, we award the top-scoring 40 % of participants Bronze, Silver and Gold certificates in the ratio 3 : 2 : 1, and we invite around 1500 of the very highest performers to take part in the Cayley, Hamilton and Maclaurin Olympiads (according to their school year) and around 12,000 to take part in either the Grey or Pink Kangaroo.

The table below shows the minimum score needed to obtain the corresponding award or follow-on round qualification for each of the previous five years of the competition.

Year of competition	2016	2017	2018	2019	2020
Maclaurin Olympiad Qual.	95	112	104	106	120
Hamilton Olympiad Qual.	89	106	100	99	117
Cayley Olympiad Qual.	81	97	93	91	110
Pink Kangaroo Qual.	66	81	76	73	89
Grey Kangaroo Qual.	60	72	68	65	80
Gold	62	78	73	71	89
Silver	50	61	59	55	72
Bronze	40	47	48	43	60
Mean score	37.1	43.2	43.4	40.0	55.2

Comments from the Problems Group

It is good to see that the Problems Group succeeded in its aim to set a more accessible paper this year. This year's mean score of 55 is significantly higher than in recent years.

The early questions were generally answered correctly by over three-quarters of the students. However, it is a little surprising that only three-quarters of them gave the correct answer to Question 1, which asked for the value of 2 - (3 - 4) - (5 - 6 - 7). Perhaps this was due to carelessness before brains were fully warmed up.

Fewer than 60% of the students answered Question 5 correctly. As the Solutions and Investigations that you can download from the UKMT website show, a very rough sketch is good enough to answer this question. It may be that in a paper where measuring instruments are not allowed, students were reluctant to draw sketches. Of course, a rough sketch might be misleading, and does not provide a proof that the correct answer is correct. However rough sketches are often useful guides, and even when rulers are allowed, a quick sketch is often as good as a diagram carefully drawn to scale.

Question 9 seemed to have caused problems, and led to several queries on the day. The question is of an unfamiliar type, but once you have understood what is being asked, finding the correct answer should not have been difficult. A sketch of the ranges on the number line would have been a great help.

Students are told that they are not expected to complete all the questions in the time available. However, we hope they are encouraged to read all the questions. Often a later question turns out to be easier than some of the questions that precede it. A case in point this year was Question 22 which was answered correctly by 27% of the students. Only 42% attempted it, but the success rate for them was 64%.

You will see from the table showing the distribution of the responses of your students to the individual questions how many of your students gave the correct answer. We hope that this information will be a guide to which questions it would be most useful to follow up in your classroom.

We congratulate your students who gained high scores, and especially those who did well enough to qualify for one of the follow-on competitions. We hope that their success will be celebrated throughout your school.

2.6 Grey Kangaroo

2.6.1 Question paper and solutions

United Kingdom
Mathematics Trust

GREY KANGAROO

Thursday 19 March 2020

a member of the Association Kangourou sans Frontières

supported by

England & Wales: Year 9 or below
Scotland: S2 or below
Northern Ireland: Year 10 or below

INSTRUCTIONS

1. Do not open the paper until the invigilator tells you to do so.
2. Time allowed: **60 minutes**.
   No answers, or personal details, may be entered after the allowed time is over.
3. The use of blank or lined paper for rough working is allowed; **squared paper, calculators and measuring instruments are forbidden**.
4. **Use a B or an HB non-propelling pencil.** Mark at most one of the options A, B, C, D, E on the Answer Sheet for each question. Do not mark more than one option.
5. **Do not expect to finish the whole paper in the time allowed.** The questions in this paper have been arranged in approximate order of difficulty with the harder questions towards the end. You are not expected to complete all the questions during the time. You should bear this in mind when deciding which questions to tackle.
6. **Scoring rules:**
   5 marks are awarded for each correct answer to Questions 1-15;
   6 marks are awarded for each correct answer to Questions 16-25;
   In this paper you will not lose marks for getting answers wrong.
7. Your Answer Sheet will be read by a machine. **Do not write or doodle on the sheet except to mark your chosen options.** The machine will read all black pencil markings even if they are in the wrong places. If you mark the sheet in the wrong place, or leave bits of eraser stuck to the page, the machine will interpret the mark in its own way.
8. **The questions on this paper are designed to challenge you to think, not to guess.** You will gain more marks, and more satisfaction, by doing one question carefully than by guessing lots of answers. This paper is about solving interesting problems, not about lucky guessing.

Enquiries about the Grey Kangaroo should be sent to:

UK Mathematics Trust, School of Mathematics, University of Leeds, Leeds LS2 9JT

☎ 0113 343 2339 enquiry@ukmt.org.uk www.ukmt.org.uk

1. Which of these fractions has the largest value?

A $\frac{8+5}{3}$ B $\frac{8}{3+5}$ C $\frac{3+5}{8}$ D $\frac{8+3}{5}$ E $\frac{3}{8+5}$

2. A large square is divided into smaller squares. In one of the smaller squares a diagonal is also drawn, as shown. What fraction of the large square is shaded?

A $\frac{4}{5}$ B $\frac{3}{8}$ C $\frac{4}{9}$ D $\frac{1}{3}$ E $\frac{1}{2}$

3. There are 4 teams in a football tournament. Each team plays every other team exactly once. In each match, the winner gets 3 points and the loser gets 0 points. In the case of a draw, both teams get 1 point. After all matches have been played, which of the following total number of points is it impossible for any team to have obtained?

A 4 B 5 C 6 D 7 E 8

4. The diagram shows a shape made up of 36 identical small equilateral triangles. What is the smallest number of small triangles identical to these that could be added to the shape to turn it into a hexagon?

A 10 B 12 C 15 D 18 E 24

5. Kanga wants to multiply three different numbers from the following list: −5, −3, −1, 2, 4, 6. What is the smallest result she could obtain?

A −200 B −120 C −90 D −48 E −15

6. John always walks to and from school at the same speed. When he walks to school along the road and walks back using a short cut across the fields, he walks for 50 minutes. When he uses the short cut both ways, he walks for 30 minutes. How long does it take him when he walks along the road both ways?

A 60 minutes B 65 minutes C 70 minutes D 75 minutes E 80 minutes

7. Each cell of a 3×3 square has a number written in it. Unfortunately the numbers are not visible because they are covered in ink. However, the sum of the numbers in each row and the sum of the numbers in two of the columns are all known, as shown by the arrows on the diagram. What is the sum of the numbers in the third column?

A 41 B 43 C 44 D 45 E 47

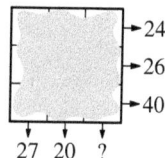

8. The shortest path from Atown to Cetown runs through Betown. The two signposts shown are set up at different places along this path. What distance is written on the broken sign?

A 1 km B 3 km C 4 km D 5 km E 9 km

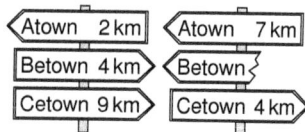

9. Anna wants to walk 5 km on average each day in March. At bedtime on 16th March, she realises that she has walked 95 km so far. What distance does she need to walk on average for the remaining days of the month to achieve her target?

A 5.4 km B 5 km C 4 km D 3.6 km E 3.1 km

10. Every pupil in a class either swims or dances. Three fifths of the class swim and three fifths dance. Five pupils both swim and dance. How many pupils are in the class?

A 15 B 20 C 25 D 30 E 35

11. Sacha's garden has the shape shown. All the sides are either parallel or perpendicular to each other. Some of the dimensions are shown in the diagram. What is the length of the perimeter of Sacha's garden?

A 22 B 23 C 24 D 25 E 26

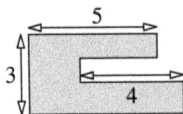

12. Werner's salary is 20% of his boss's salary. By what percentage is his boss's salary larger than Werner's salary?

A 80% B 120% C 180% D 400% E 520%

13. The pattern on a large square tile consists of eight congruent right-angled triangles and a small square. The area of the tile is 49 cm^2 and the length of the hypotenuse PQ of one of the triangles is 5 cm. What is the area of the small square?

A 1 cm^2 B 4 cm^2 C 9 cm^2 D 16 cm^2 E 25 cm^2

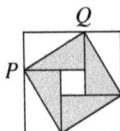

14. Andrew buys 27 identical small cubes, each with two adjacent faces painted red. He then uses all of these cubes to build a large cube. What is the largest number of completely red faces that the large cube can have?

A 2 B 3 C 4 D 5 E 6

15. Aisha has a strip of paper with the numbers 1, 2, 3, 4 and 5 written in five cells as shown. She folds the strip so that the cells overlap, forming 5 layers. Which of the following configurations, from top layer to bottom layer, is it not possible to obtain?

1	2	3	4	5

A 3, 5, 4, 2, 1 B 3, 4, 5, 1, 2 C 3, 2, 1, 4, 5 D 3, 1, 2, 4, 5 E 3, 4, 2, 1, 5

16. Twelve coloured cubes are arranged in a row. There are 3 blue cubes, 2 yellow cubes, 3 red cubes and 4 green cubes but not in that order. There is a yellow cube at one end and a red cube at the other end. The red cubes are all together within the row. The green cubes are also all together within the row. The tenth cube from the left is blue. What colour is the cube sixth from the left?

A green B yellow C blue D red E red or blue

17. Bella took a square piece of paper and folded two of its sides to lie along the diagonal, as shown, to obtain a quadrilateral. What is the largest size of an angle in that quadrilateral?

A 112.5° B 120° C 125° D 135° E 150°

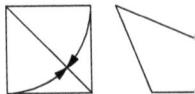

18. How many four-digit numbers N are there, such that half of the number N is divisible by 2, a third of N is divisible by 3 and a fifth of N is divisible by 5?

A 1 B 7 C 9 D 10 E 11

19. In the final of a dancing competition, each of the three members of the jury gives each of the five competitors 0 points, 1 point, 2 points, 3 points or 4 points. No two competitors get the same mark from any individual judge. Adam knows all the sums of the marks and a few single marks, as shown. How many points does Adam get from judge III?

	Adam	Berta	Clara	David	Emil
I	2	0			
II		2	0		
III					
Sum	7	5	3	4	11

A 0 B 1 C 2 D 3 E 4

20. Harriet writes a positive integer on each edge of a square. She also writes at each vertex the product of the integers on the two edges that meet at that vertex. The sum of the integers at the vertices is 15. What is the sum of the integers on the edges of the square?

A 6 B 7 C 8 D 10 E 15

21. Sophia has 52 identical isosceles right-angled triangles. She wants to make a square using some of them. How many different-sized squares could she make?

A 6 B 7 C 8 D 9 E 10

22. Cleo builds a pyramid with identical metal spheres. Its square base is a 4×4 array of spheres, as shown in the diagram. The upper layers are a 3×3 array of spheres, a 2×2 array of spheres and a single sphere at the top. At each point of contact between two spheres, a blob of glue is placed. How many blobs of glue will Cleo place?

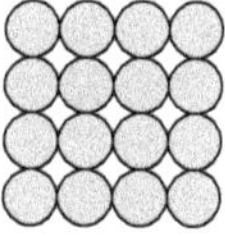

A 72 B 85 C 88 D 92 E 96

23. Four children are in the four corners of a 10 m × 25 m pool. Their coach is standing somewhere on one side of the pool. When he calls them, three children get out and walk as short a distance as possible round the pool to meet him. They walk 50 m in total. What is the shortest distance the coach needs to walk to get to the fourth child's corner?

A 10 m B 12 m C 15 m D 20 m E 25 m

24. Anne, Bronwyn and Carl ran a race. They started at the same time, and their speeds were constant. When Anne finished, Bronwyn had 15 m to run and Carl had 35 m to run. When Bronwyn finished, Carl had 22 m to run. What was the length of the race?

A 135 m B 140 m C 150 m D 165 m E 175 m

25. The statements on the right give clues to the identity of a four-digit number.

What is the last digit of the four-digit number?

A 0 B 1 C 3
D 5 E 9

Number	Clue
4 1 3 2	Two digits are correct but in the wrong places.
9 8 2 6	One digit is correct and in the right place.
5 0 7 9	Two digits are correct with one of them being in the right place and the other one in the wrong place.
2 7 4 1	One digit is correct but in the wrong place.
7 6 4 2	None of the digits is correct.

1. Which of these fractions has the largest value?

A $\frac{8+5}{3}$ B $\frac{8}{3+5}$ C $\frac{3+5}{8}$ D $\frac{8+3}{5}$ E $\frac{3}{8+5}$

SOLUTION **A**

The values of the fractions shown are $\frac{13}{3} = 4\frac{1}{3}$, $\frac{8}{8} = 1$, $\frac{8}{8} = 1$, $\frac{11}{5} = 2\frac{1}{5}$ and $\frac{3}{13}$. Hence the fraction which has the largest value is $\frac{8+5}{3}$.

2. A large square is divided into smaller squares. In one of the smaller squares a diagonal is also drawn, as shown. What fraction of the large square is shaded?

A $\frac{4}{5}$ B $\frac{3}{8}$ C $\frac{4}{9}$ D $\frac{1}{3}$ E $\frac{1}{2}$

SOLUTION **E**

The shaded square in the lower right corner of the large square is $\frac{1}{4}$ of the large square. The shaded triangle is half of $\frac{1}{4}$ of the large square. Hence it is $\frac{1}{8}$ of the large square. The two small shaded squares in the upper left corner together are half of $\frac{1}{4}$, or $\frac{1}{8}$, of the large square. Therefore the fraction of the large square that is shaded is $\frac{1}{4} + \frac{1}{8} + \frac{1}{8} = \frac{1}{2}$.

3. There are 4 teams in a football tournament. Each team plays every other team exactly once. In each match, the winner gets 3 points and the loser gets 0 points. In the case of a draw, both teams get 1 point. After all matches have been played, which of the following total number of points is it impossible for any team to have obtained?

A 4 B 5 C 6 D 7 E 8

SOLUTION **E**

Each team plays exactly three matches. Hence the maximum number of points any team can obtain is $3 \times 3 = 9$. A draw only gets 1 point. Hence the next highest total number of points possible, from two wins and a draw, is $2 \times 3 + 1 = 7$. Therefore it is impossible to obtain 8 points.
(Note: totals of 4, 5 and 6 points can be obtained by one win, one draw and one loss, one win and two draws and two wins and a loss respectively.)

4. The diagram shows a shape made up of 36 identical small equilateral triangles. What is the smallest number of small triangles identical to these that could be added to the shape to turn it into a hexagon?

A 10 B 12 C 15 D 18 E 24

Solution **D**

To turn the figure given in the question into a hexagon by adding the smallest number of triangles, two triangles should be added to create each vertex of the hexagon (shaded dark grey) and one triangle added to create each edge (shaded light grey) as shown on the right for one edge. Since a hexagon has six vertices and six edges, the smallest number of triangles required is $6 \times (2 + 1) = 18$.

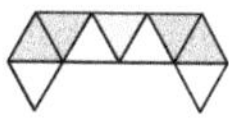

5. Kanga wants to multiply three different numbers from the following list: $-5, -3, -1, 2, 4, 6$. What is the smallest result she could obtain?

A -200 B -120 C -90 D -48 E -15

Solution **B**

The result furthest from zero is obtained by multiplying the three numbers furthest from zero, namely -5, 4 and 6. This gives -120, which is negative and hence is the smallest possible result.

6. John always walks to and from school at the same speed. When he walks to school along the road and walks back using a short cut across the fields, he walks for 50 minutes. When he uses the short cut both ways, he walks for 30 minutes. How long does it take him when he walks along the road both ways?

A 60 minutes B 65 minutes C 70 minutes D 75 minutes E 80 minutes

Solution **C**

Since using the short cut both ways takes 30 minutes, using the short cut one way takes 15 minutes. Hence, since walking to school by road and walking back using the short cut takes 50 minutes, walking by road takes $(50 - 15)$ minutes $= 35$ minutes. Therefore walking by road both ways takes 2×35 minutes $= 70$ minutes.

7. Each cell of a 3×3 square has a number written in it. Unfortunately the numbers are not visible because they are covered in ink. However, the sum of the numbers in each row and the sum of the numbers in two of the columns are all known, as shown by the arrows on the diagram. What is the sum of the numbers in the third column?

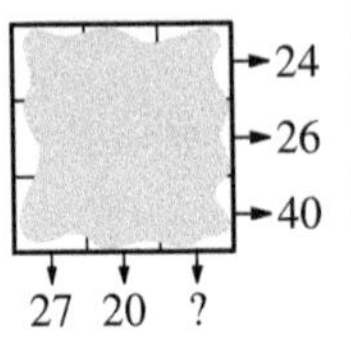

A 41 B 43 C 44 D 45 E 47

SOLUTION **B**

The sum of the row totals is the sum of all the nine numbers in the 3×3 square. Likewise, the sum of the column totals is the sum of these nine numbers. Therefore $24+26+40 = 27+20+x$, where x is the sum of the numbers in the third column. Therefore $90 = 47 + x$ and hence $x = 43$. Therefore the sum of the numbers in the third column is 43.

8. The shortest path from Atown to Cetown runs through Betown. The two signposts shown are set up at different places along this path. What distance is written on the broken sign?

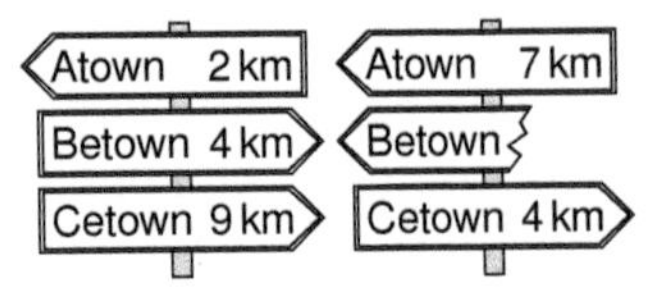

A 1 km B 3 km C 4 km D 5 km E 9 km

SOLUTION **A**

The information on the signs pointing to Atown and the signs pointing to Cetown both tell us that the distance between the signs is $(7 - 2)$ km $= (9 - 4)$ km $= 5$ km. Therefore the distance which is written on the broken sign is $(5 - 4)$ km $= 1$ km.

9. Anna wants to walk 5 km on average each day in March. At bedtime on 16th March, she realises that she has walked 95 km so far. What distance does she need to walk on average for the remaining days of the month to achieve her target?

A 5.4 km B 5 km C 4 km D 3.6 km E 3.1 km

SOLUTION **C**

The total distance Anna wants to walk is (31×5) km $= 155$ km. Since she has walked 95 km up to the 16th of March, she has $(155 - 95)$ km $= 60$ km to walk in $(31 - 16)$ days $= 15$ days. Therefore the distance, in km, that she needs to average per day is $60 \div 15 = 4$.

> **10.** Every pupil in a class either swims or dances. Three fifths of the class swim and three fifths dance. Five pupils both swim and dance. How many pupils are in the class?
>
> A 15 B 20 C 25 D 30 E 35

SOLUTION  **C**

Since three fifths of the class swim, three fifths of the class dance and no-one does neither, the fraction of the class who do both is $\frac{3}{5} + \frac{3}{5} - 1 = \frac{1}{5}$. Hence, since 5 pupils do both, the number of pupils in the class is $5 \times 5 = 25$.

> **11.** Sacha's garden has the shape shown. All the sides are either parallel or perpendicular to each other. Some of the dimensions are shown in the diagram. What is the length of the perimeter of Sacha's garden?
>
> 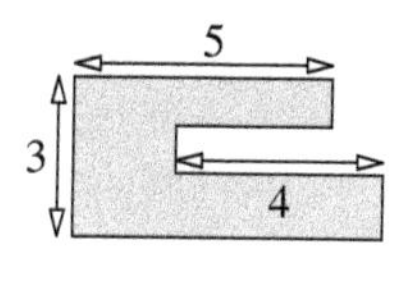
>
>
> 
> A 22 B 23 C 24 D 25 E 26

SOLUTION **C**

Divide the garden up and let the lengths of the various sides be as shown on the diagram. Since all sides are either parallel or perpendicular, $a + b + c = 3$. Therefore the perimeter of Sacha's garden is $3 + 5 + a + x + b + 4 + c + (4 + (5 - x))$ $= 21 + a + b + c = 24$.

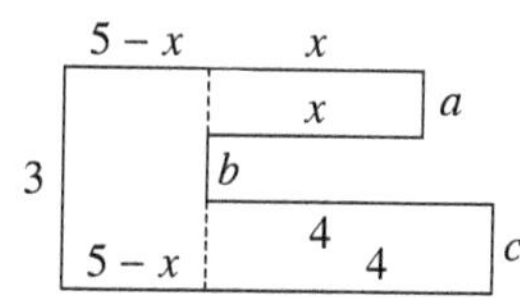

> **12.** Werner's salary is 20% of his boss's salary. By what percentage is his boss's salary larger than Werner's salary?
>
> A 80% B 120% C 180% D 400% E 520%

SOLUTION **D**

Werner's salary is 20% of his boss's salary. Therefore his boss's salary is $\frac{100}{20} = 5$ times Werner's salary. Hence his boss's salary is 500% of Werner's salary and so is 400% larger.

13. The pattern on a large square tile consists of eight congruent right-angled triangles and a small square. The area of the tile is 49 cm^2 and the length of the hypotenuse PQ of one of the triangles is 5 cm. What is the area of the small square?

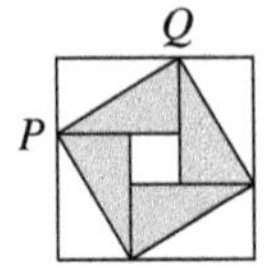

A 1 cm^2 B 4 cm^2 C 9 cm^2 D 16 cm^2 E 25 cm^2

SOLUTION **A**

Since the four rectangles are congruent, the diagonal PQ is also the side of a square. This square has area $(5 \times 5)\,\text{cm}^2 = 25\,\text{cm}^2$. Therefore the total area of the rectangles outside the square with side PQ but inside the large square is $(49 - 25)\,\text{cm}^2 = 24\,\text{cm}^2$. However, this is also equal to the total area of the triangles inside the square with side PQ. Therefore the area of the small square is $(25 - 24)\,\text{cm}^2 = 1\,\text{cm}^2$.

14. Andrew buys 27 identical small cubes, each with two adjacent faces painted red. He then uses all of these cubes to build a large cube. What is the largest number of completely red faces that the large cube can have?

A 2 B 3 C 4 D 5 E 6

SOLUTION **C**

Since no small cubes have three faces painted red, it is impossible to build a large cube with three faces that meet at a vertex painted red. Therefore no more than four of the faces can be red and, without loss of generality, let us consider that they are the front, back and sides of the cube. We can construct a large cube with four faces completely red by arranging that the 12 small cubes along the four vertical edges of the large cube have two red faces showing and the 12 small cubes down the centre of the four vertical faces have one red face showing. Hence it is possible to build a large cube with four completely red faces.

15. Aisha has a strip of paper with the numbers 1, 2, 3, 4 and 5 written in five cells as shown. She folds the strip so that the cells overlap, forming 5 layers. Which of the following configurations, from top layer to bottom layer, is it not possible to obtain?

1	2	3	4	5

A 3, 5, 4, 2, 1 B 3, 4, 5, 1, 2 C 3, 2, 1, 4, 5 D 3, 1, 2, 4, 5
E 3, 4, 2, 1, 5

SOLUTION **E**

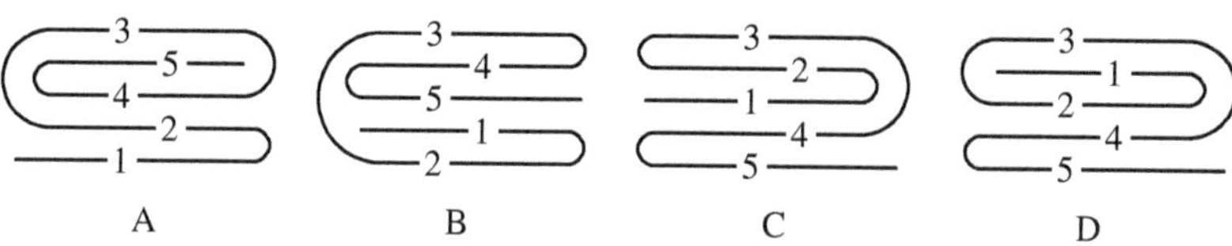

The four figures (A) to (D) give a side-view of how the strip could be folded to give the arrangements of numbers in options A to D. Figure (E) shows that it is not possible to get option E since number 5 would end up between number 4 and number 2 (as indicated by the dashed line labelled 5) rather than below number 1 as is required.

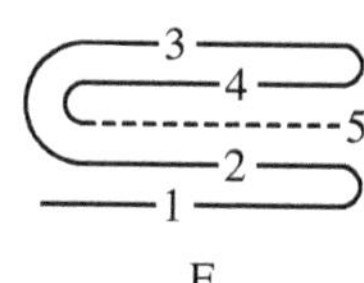

16. Twelve coloured cubes are arranged in a row. There are 3 blue cubes, 2 yellow cubes, 3 red cubes and 4 green cubes but not in that order. There is a yellow cube at one end and a red cube at the other end. The red cubes are all together within the row. The green cubes are also all together within the row. The tenth cube from the left is blue. What colour is the cube sixth from the left?

A green B yellow C blue D red
E red or blue

SOLUTION **A**

We are told there is a red cube at one end and that the three red cubes are all together within the row. Therefore there is a block of three red cubes at one end. If they were at the right-hand end, the tenth cube from the left would be red. But the tenth cube from the left is blue. Hence the red cubes are at the left-hand end and a yellow cube is at the right-hand end, with a blue cube in the tenth place from the left, as shown in the diagram.

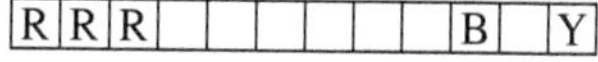

The four green cubes are all together within the row and hence are somewhere between the 4th and 9th positions from the left. Whether they start at position 4 or 5 or 6 from the left, the cube in the 6th position from the left is green.

17. Bella took a square piece of paper and folded two of its sides to lie along the diagonal, as shown, to obtain a quadrilateral. What is the largest size of an angle in that quadrilateral?

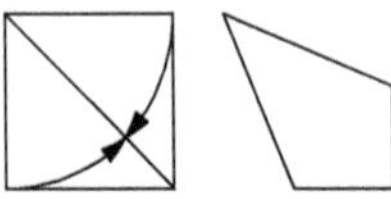

A 112.5°   B 120°   C 125°   D 135°   E 150°

SOLUTION **A**

Since the quadrilateral is formed by folding the 45° angles above and below the diagonal of the square in half, the size of the small angle of the quadrilateral is $2 \times (\frac{1}{2} \times 45°) = 45°$. One angle of the quadrilateral is 90° and the other two are equal from the construction. Therefore, since the sum of the angles in a quadrilateral is 360°, the size of the equal angles is $(360 - 90 - 45)° \div 2 = 225° \div 2 = 112.5°$.

18. How many four-digit numbers N are there, such that half of the number N is divisible by 2, a third of N is divisible by 3 and a fifth of N is divisible by 5?

A 1   B 7   C 9   D 10   E 11

SOLUTION **D**

The information in the question tells us that N is divisible by $2 \times 2 = 4$, $3 \times 3 = 9$ and $5 \times 5 = 25$. Since these three values have no factors in common, N is divisible by $4 \times 9 \times 25 = 900$. Therefore N is a four-digit multiple of 900. The smallest such multiple is $2 \times 900 = 1800$ and the largest is $11 \times 900 = 9900$. Therefore there are 10 such four-digit numbers.

19. In the final of a dancing competition, each of the three members of the jury gives each of the five competitors 0 points, 1 point, 2 points, 3 points or 4 points. No two competitors get the same mark from any individual judge. Adam knows all the sums of the marks and a few single marks, as shown. How many points does Adam get from judge III?

	Adam	Berta	Clara	David	Emil
I	2	0			
II		2	0		
III					
Sum	7	5	3	4	11

A 0 B 1 C 2 D 3 E 4

Solution **B**

The table can be partially completed as follows. Berta scored 5 points in total. Therefore her score from judge III is 3. Clara's total is 3. Therefore she cannot have been given 4 by any of the judges. David's total is 4. He cannot have been given a score of 4 by any judge, since, if so, both the other two judges must have given him 0. This is impossible, as judges I and II give 0 to Berta and Clara, respectively. Therefore judge I gives 4 to Emil, and judge II gives 4 to either Adam or Emil. Emil's total is 11. So if he gets 4 from judges I and II, he gets 3 from judge III which is not possible as judge III gave Berta 3. Hence judge II gives 4 to Adam. Because Adam's total is 7, it now follows that Adam gets a score of 1 from judge III.
(Note: the final four scores are not uniquely determined.)

	Adam	Berta	Clara	David	Emil
I	2	0			(4)
II	(4)	2	0	(1)	(3)
III	(1)	(3)			(4)
Sum	7	5	3	4	11

20. Harriet writes a positive integer on each edge of a square. She also writes at each vertex the product of the integers on the two edges that meet at that vertex. The sum of the integers at the vertices is 15. What is the sum of the integers on the edges of the square?

A 6 B 7 C 8 D 10 E 15

SOLUTION **C**

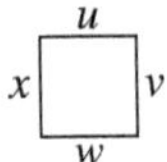

Let the integers written on the edges of the square be u, v, w and x as shown in the diagram. Therefore, since the integer written at each vertex is the product of the integers written on the edges that meet at that vertex, the integers at the vertices are ux, uv, vw and wx. Therefore $ux + uv + vw + wx = 15$ and hence $(u + w)(v + x) = 15$. Since u, v, w and x are positive integers, the smallest either of $u + w$ or $v + x$ can be is $1 + 1 = 2$. Therefore the only possible multiplications which give an answer of 15 are $3 \times 5 = 15$ and $5 \times 3 = 15$. In each case, the sum of the integers on the edges of the square is $3 + 5 = 8$.

21. Sophia has 52 identical isosceles right-angled triangles. She wants to make a square using some of them. How many different-sized squares could she make?

A 6 B 7 C 8 D 9 E 10

SOLUTION **C**

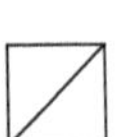
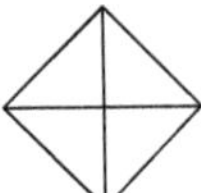

Sophia can make a small square by joining two of her isosceles right-angled triangles, as shown in the first diagram. She can then join 4 or 9 or 16 or 25 of these small squares to make larger squares using up to 50 triangles. She can also create a different small square by joining four of her isosceles triangles, as shown in the second diagram. Note that the side-length of this square is $\sqrt{2}$ times the side-length of the previous small square. She can then join 4 or 9 of this second type of small square to make larger squares using up to 36 triangles. The eight squares described so far all have different side-lengths and it can be shown that there are no other possible sizes of squares. Hence Sophia can make 8 different-sized squares with the triangles she has.

22. Cleo builds a pyramid with identical metal spheres. Its square base is a 4×4 array of spheres, as shown in the diagram. The upper layers are a 3×3 array of spheres, a 2×2 array of spheres and a single sphere at the top. At each point of contact between two spheres, a blob of glue is placed. How many blobs of glue will Cleo place?

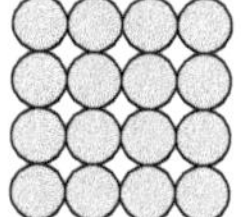

A 72 B 85 C 88 D 92 E 96

SOLUTION **E**

Consider first how to join each individual layer. In the 4×4 layer there are $4 \times 3 \times 2 = 24$ points of contact. Similarly, in the 3×3 layer there are $3 \times 2 \times 2 = 12$ points of contact and in the 2×2 layer there are $2 \times 1 \times 2 = 4$ points of contact. Now consider how to join the layers together. To join the 4×4 layer to the 3×3 layer, each sphere in the 3×3 layer has 4 points of contact with the lower layer, making $3 \times 3 \times 4 = 36$ points of contact. Similarly to join the 3×3 layer to the 2×2 layer, each sphere in the 2×2 layer has 4 points of contact with the lower layer making $2 \times 2 \times 4 = 16$ points of contact and the single sphere in the top layer has 4 points of contact with the 2×2 layer. Hence the total number of points of contact is $24 + 12 + 4 + 36 + 16 + 4 = 96$ and therefore 96 blobs of glue are used.

23. Four children are in the four corners of a 10 m × 25 m pool. Their coach is standing somewhere on one side of the pool. When he calls them, three children get out and walk as short a distance as possible round the pool to meet him. They walk 50 m in total. What is the shortest distance the coach needs to walk to get to the fourth child's corner?

A 10 m B 12 m C 15 m D 20 m E 25 m

SOLUTION **D**

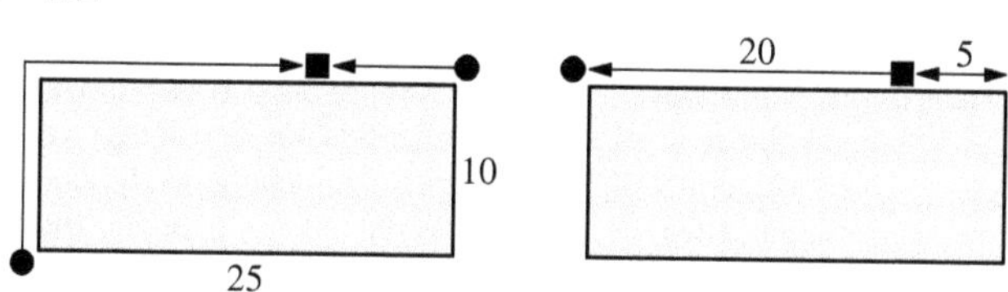

Consider two children in opposite corners of the pool. Wherever their trainer stands, the total distance these two pupils would need to walk to meet him is half the perimeter of the pool, as illustrated in the first diagram. Therefore, if all four children walked to meet their trainer, the total distance they would walk is $(2 \times 10 + 2 \times 25)$ m = 70 m. Since we are given that three of the children walked 50 m in total, the distance the trainer would have to walk to get to the fourth child is $(70 - 50)$ m = 20 m. The second diagram shows one possible position of the trainer and child which satisfies this situation although there are others.

24. Anne, Bronwyn and Carl ran a race. They started at the same time, and their speeds were constant. When Anne finished, Bronwyn had 15 m to run and Carl had 35 m to run. When Bronwyn finished, Carl had 22 m to run. What was the length of the race?

A 135 m B 140 m C 150 m D 165 m E 175 m

SOLUTION **D**

First note that Carl ran $(35 - 22)$ m $= 13$ m while Bronwen ran 15 m. Let the length of the race be x m. Since their speeds were constant, the ratio of the distances they ran in any time is also constant. Therefore, since Carl ran $(x - 35)$ m while Bronwen ran $(x - 15)$ m, we have $\frac{x-35}{x-15} = \frac{13}{15}$. Therefore $15x - 15 \times 35 = 13x - 13 \times 15$ and hence $2x = 15 \times 35 - 13 \times 15$. Therefore $2x = 22 \times 15$ and hence $x = 11 \times 15$. Therefore the distance they ran is 165 m.

25.

4 1 3 2	Two digits are correct but in the wrong places.
9 8 2 6	One digit is correct and in the right place.
5 0 7 9	Two digits are correct with one of them being in the right place and the other one in the wrong place.
2 7 4 1	One digit is correct but in the wrong place.
7 6 4 2	None of the digits is correct.

The statements above give clues to the identity of a four-digit number.

What is the last digit of the four-digit number?

A 0 B 1 C 3 D 5 E 9

SOLUTION **C**

Let's call the four-digit number N. The last clue tells us that none of the digits 7, 6, 4 or 2 is a digit in N. Then the fourth clue shows that 1 is a digit in N, but it is not the fourth digit. The first clue now tells us that N involves a 3 but not as its third digit. It also shows that 1 is not the second digit. The second clue now tells us that either 8 is the second digit of N and 9 is not one of its digits or else 9 is the first digit of N and 8 is not one of its digits. Suppose that 8 were the correct second digit. Then the third clue would tell us that both 0 and 5 were correct digits. But this would mean that all of 1, 3, 8, 0 and 5 were digits of the four-digit number N. Therefore 8 is incorrect and so 9 is correct and is the first digit. Knowing this, the third clue shows us that exactly one of 5 and 0 is correct and, moreover, it is in the right place. It can't be 5 because the first digit of N is 9. So 0 is the correct second digit. We already know 3 is correct, but is not the third digit; so the last digit of N is 3 and N is 9013.

2.6.2 Participation

Participation in the Grey Kangaroo

Year of competition	2017	2018	2019	2020
Centres with invitees	1,193	1,332	1,126	628
Centres returning scripts	1,036	1,146	1,104	619
Invitees	3,862	4,461	3,866	4903
Scripts returned	3,146	3,512	3,271	1877

Percentage change from previous year

Year of competition	2017	2018	2019	2020
Centres with invitees	-4.3%	+11.7%	-15.7%	-44.2%
Centres returning scripts	+0.7%	+10.6%	-3.7%	-43.9%
Invitees	+17.4%	+15.5%	-13.3%	+26.8%
Scripts returned	+24.6%	+11.6%	-6.9%	-42.7%

2.6.3 Student performance

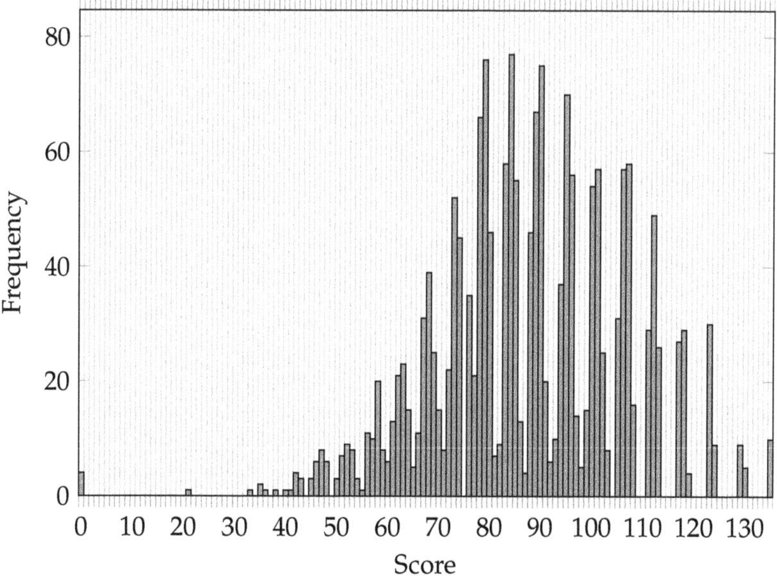

Answers given

The table below shows, for each question, the percentage of students giving each answer. The correct answer is underlined.

Question	A	B	C	D	E	Blank	Ambiguous
1	98	1	0	1	0	0	0
2	0	1	0	0	98	0	0
3	4	3	3	3	84	2	0
4	1	5	4	86	4	1	0
5	1	93	2	0	4	1	0
6	1	0	96	1	1	0	0
7	5	72	6	7	4	5	0
8	88	3	7	6	7	1	0
9	1	4	85	6	2	1	0
10	1	2	90	5	2	1	0
11	11	8	65	7	5	3	0
12	10	7	4	70	8	1	0
13	66	16	10	3	3	2	0
14	19	13	49	5	12	2	0
15	11	13	5	9	59	3	0
16	92	1	3	1	2	2	0
17	46	14	19	11	3	7	0
18	12	15	20	30	10	13	0
19	3	76	5	3	9	4	0
20	4	10	48	14	10	14	0
21	30	16	19	10	8	17	0
22	14	8	9	12	44	12	1
23	10	12	17	37	7	16	0
24	19	16	20	28	6	21	0
25	7	8	46	10	14	14	0

Mean scores and award boundaries

The table shows, for each of the past five years of the competition, the minimum score needed to obtain a Certificate of Merit and the mean score.

Year of competition	2017	2018	2019	2020
Certificate of Merit	100	93	107	101
Mean score	85.8	80.6	93.8	87.4

2.7 Pink Kangaroo

2.7.1 Question paper and solutions

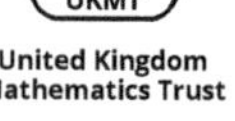

PINK KANGAROO

Thursday 19 March 2020

a member of the Association Kangourou sans Frontières

supported by 

England & Wales: Year 11 or below
Scotland: S4 or below
Northern Ireland: Year 12 or below

INSTRUCTIONS

1. Do not open the paper until the invigilator tells you to do so.
2. Time allowed: **60 minutes**.
   No answers, or personal details, may be entered after the allowed time is over.
3. The use of blank or lined paper for rough working is allowed; **squared paper**, **calculators and measuring instruments are forbidden**.
4. **Use a B or an HB non-propelling pencil.** Mark at most one of the options A, B, C, D, E on the Answer Sheet for each question. Do not mark more than one option.
5. **Do not expect to finish the whole paper in the time allowed.** The questions in this paper have been arranged in approximate order of difficulty with the harder questions towards the end. You are not expected to complete all the questions during the time. You should bear this in mind when deciding which questions to tackle.
6. **Scoring rules:**
   5 marks are awarded for each correct answer to Questions 1-15;
   6 marks are awarded for each correct answer to Questions 16-25;
   In this paper you will not lose marks for getting answers wrong.
7. Your Answer Sheet will be read by a machine. **Do not write or doodle on the sheet except to mark your chosen options.** The machine will read all black pencil markings even if they are in the wrong places. If you mark the sheet in the wrong place, or leave bits of eraser stuck to the page, the machine will interpret the mark in its own way.
8. **The questions on this paper are designed to challenge you to think, not to guess.** You will gain more marks, and more satisfaction, by doing one question carefully than by guessing lots of answers. This paper is about solving interesting problems, not about lucky guessing.

Enquiries about the Pink Kangaroo should be sent to:

UK Mathematics Trust, School of Mathematics, University of Leeds, Leeds LS2 9JT

☎ 0113 343 2339 enquiry@ukmt.org.uk www.ukmt.org.uk

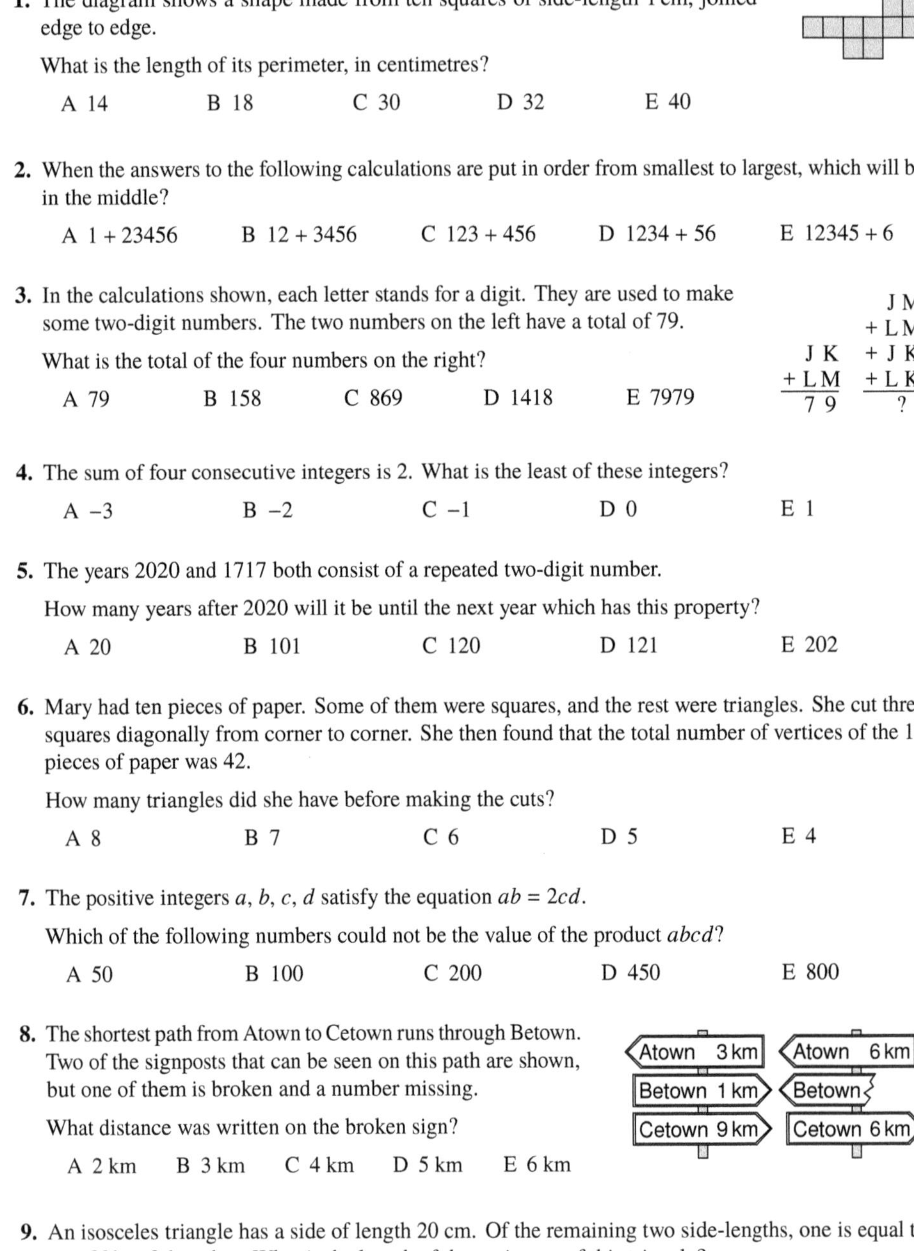

1. The diagram shows a shape made from ten squares of side-length 1 cm, joined edge to edge.

What is the length of its perimeter, in centimetres?

A 14 B 18 C 30 D 32 E 40

2. When the answers to the following calculations are put in order from smallest to largest, which will be in the middle?

A $1 + 23456$ B $12 + 3456$ C $123 + 456$ D $1234 + 56$ E $12345 + 6$

3. In the calculations shown, each letter stands for a digit. They are used to make some two-digit numbers. The two numbers on the left have a total of 79.

What is the total of the four numbers on the right?

A 79 B 158 C 869 D 1418 E 7979

$$\begin{array}{r} \mathrm{J\,K} \\ +\,\mathrm{L\,M} \\ \hline 7\,9 \end{array} \qquad \begin{array}{r} \mathrm{J\,M} \\ +\,\mathrm{L\,M} \\ +\,\mathrm{J\,K} \\ +\,\mathrm{L\,K} \\ \hline ? \end{array}$$

4. The sum of four consecutive integers is 2. What is the least of these integers?

A -3 B -2 C -1 D 0 E 1

5. The years 2020 and 1717 both consist of a repeated two-digit number.

How many years after 2020 will it be until the next year which has this property?

A 20 B 101 C 120 D 121 E 202

6. Mary had ten pieces of paper. Some of them were squares, and the rest were triangles. She cut three squares diagonally from corner to corner. She then found that the total number of vertices of the 13 pieces of paper was 42.

How many triangles did she have before making the cuts?

A 8 B 7 C 6 D 5 E 4

7. The positive integers a, b, c, d satisfy the equation $ab = 2cd$.

Which of the following numbers could not be the value of the product $abcd$?

A 50 B 100 C 200 D 450 E 800

8. The shortest path from Atown to Cetown runs through Betown. Two of the signposts that can be seen on this path are shown, but one of them is broken and a number missing.

What distance was written on the broken sign?

A 2 km B 3 km C 4 km D 5 km E 6 km

9. An isosceles triangle has a side of length 20 cm. Of the remaining two side-lengths, one is equal t two-fifths of the other. What is the length of the perimeter of this triangle?

A 36 cm B 48 cm C 60 cm D 90 cm E 120 cm

10. Freda wants to write a number in each of the nine cells of this figure so that the sum of the three numbers on each diameter is 13 and the sum of the eight numbers on the circumference is 40.

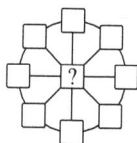

What number must be written in the central cell?

A 3 B 5 C 8 D 10 E 12

11. Masha put a multiplication sign between the second and third digits of the number 2020 and noted that the resulting product 20×20 was a square number.

How many integers between 2010 and 2099 (including 2020) have the same property?

A 1 B 2 C 3 D 4 E 5

12. Two squares of different sizes are drawn inside an equilateral triangle. One side of one of these squares lies on one of the sides of the triangle as shown. What is the size of the angle marked by the question mark?

A 25° B 30° C 35° D 45° E 50°

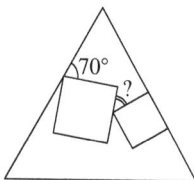

13. Luca began a 520 km trip by car with 14 litres of fuel in the car tank. His car consumes 1 litre of fuel per 10 km. After driving 55 km, he saw a road sign showing the distances from that point to five petrol stations ahead on the road. These distances are 35 km, 45 km, 55 km, 75 km and 95 km. The capacity of the car's fuel tank is 40 litres and Luca wants to stop just once to fill the tank.

How far is the petrol station that he should stop at?

A 35 km B 45 km C 55 km D 75 km E 95 km

14. The numbers x and y satisfy the equation $17x + 51y = 102$. What is the value of $9x + 27y$?

A 54 B 36 C 34 D 18
E The value is undetermined.

15. A vertical stained glass square window of area 81 cm^2 is made out of six triangles of equal area (see figure). A fly is sitting on the exact spot where the six triangles meet. How far from the bottom of the window is the fly sitting?

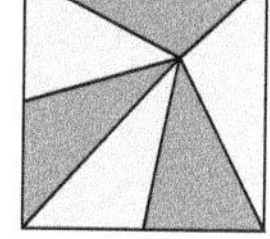

A 3 cm B 5 cm C 5.5 cm D 6 cm E 7.5 cm

16. The digits from 1 to 9 are randomly arranged to make a 9-digit number.

What is the probability that the resulting number is divisible by 18?

A $\frac{1}{3}$ B $\frac{4}{9}$ C $\frac{1}{2}$ D $\frac{5}{9}$ E $\frac{3}{4}$

17. A hare and a tortoise competed in a 5 km race along a straight line, going due North. The hare is five times as fast as the tortoise. The hare mistakenly started running due East. After a while he realised his mistake, then turned and ran straight to the finish point. He arrived at the same time as the tortoise. What was the distance between the hare's turning point and the finish point?

A 11 km B 12 km C 13 km D 14 km E 15 km

18. There are some squares and triangles on the table. Some of them are blue and the rest are red. Some of these shapes are large and the rest are small. We know that

1. If the shape is large, it's a square;
2. If the shape is blue, it's a triangle.

Which of the statements A–E must be true?

A All red figures are squares. B All squares are large. C All small figures are blue.
D All triangles are blue. E All blue figures are small.

19. Two identical rectangles with sides of length 3 cm and 9 cm are overlapping as in the diagram. What is the area of the overlap of the two rectangles?

A 12 cm^2 B 13.5 cm^2 C 14 cm^2 D 15 cm^2 E 16 cm^2

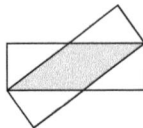

20. Kanga labelled the vertices of a square-based pyramid using 1, 2, 3, 4 and 5 once each. For each face Kanga calculated the sum of the numbers on its vertices. Four of these sums equalled 7, 8, 9 and 10. What is the sum for the fifth face?

A 11 B 12 C 13 D 14 E 15

21. A large cube is built using 64 smaller identical cubes. Three of the faces of the large cube are painted. What is the maximum possible number of small cubes that can have exactly one face painted?

A 27 B 28 C 32 D 34 E 40

22. In each of the cells, a number is to be written so that the sum of the 4 numbers in each row and in each column are the same.

What number must be written in the shaded cell?

A 5 B 6 C 7 D 8 E 9

1		6	3
	2	2	8
	7		4
(shaded)		7	

23. Alice, Belle and Cathy had an arm-wrestling contest. In each game two girls wrestled, while the third rested. After each game, the winner played the next game against the girl who had rested. In total Alice played 10 times, Belle played 15 times and Cathy played 17 times. Who lost the second game?

A Alice
B Belle
C Cathy
D Either Alice or Belle could have lost the second game.
E Either Belle or Cathy could have lost the second game.

24. Eight consecutive three-digit positive integers have the following property: each of them is divisible by its last digit. What is the sum of the digits of the smallest of these eight integers?

A 9 B 10 C 11 D 12 E 13

25. A zig-zag line starts at the point P, at one end of the diameter PQ of a circle. Each of the angles between the zig-zag line and the diameter PQ is equal to α as shown. After four peaks, the zig-zag line ends at the point Q. What is the size of angle α?

A 60° B 72° C 75° D 80° E 86°

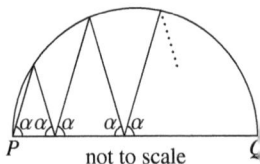

---

1. The diagram shows a shape made from ten squares of side-length 1 cm, joined edge to edge.

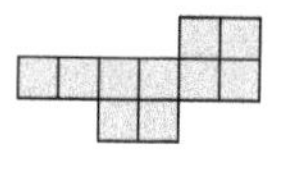

What is the length of its perimeter, in centimetres?

A 14 B 18 C 30 D 32 E 40

SOLUTION **B**

Counting the edges of the squares around the shape gives a perimeter of 18 cm.

2. When the answers to the following calculations are put in order from smallest to largest, which will be in the middle?

A $1 + 23456$ B $12 + 3456$ C $123 + 456$ D $1234 + 56$ E $12345 + 6$

SOLUTION **B**

The answers are 23457, 3468, 579, 1290 and 12351 respectively, so in ascending order the middle one is B.

3. In the calculations shown, each letter stands for a digit. They are used to make some two-digit numbers. The two numbers on the left have a total of 79.

What is the total of the four numbers on the right?

$$\begin{array}{r} \text{J K} \\ +\,\text{L M} \\ \hline 7\ 9 \end{array} \qquad \begin{array}{r} \text{J M} \\ +\,\text{L M} \\ +\,\text{J K} \\ +\,\text{L K} \\ \hline ? \end{array}$$

A 79 B 158 C 869 D 1418 E 7979

SOLUTION **B**

The numbers on the left use the digits M and K in the Units column, and the digits J and L in the Tens column. The numbers on the right use the digits M and K twice in the Units columns, and the digits J and L twice in the Tens column, so the total is exactly twice that of the numbers on the left. Twice 79 is 158, answer B.

4. The sum of four consecutive integers is 2. What is the least of these integers?

A -3 B -2 C -1 D 0 E 1

SOLUTION **C**

The four integers cannot all be non-positive since the total is positive. Also, they cannot all be non-negative since the smallest possible total would then be $0 + 1 + 2 + 3 = 6$. So, they must have at least one negative integer, and at least one positive integer, and hence will include -1, 0 and 1. Since these have a total of 0, the fourth number must be 2. Thus the least integer is -1.

5. The years 2020 and 1717 both consist of a repeated two-digit number.

How many years after 2020 will it be until the next year which has this property?

A 20 B 101 C 120 D 121 E 202

SOLUTION **B**

The next year with this property is 2121 which is 101 years after 2020.

6. Mary had ten pieces of paper. Some of them were squares, and the rest were triangles. She cut three squares diagonally from corner to corner. She then found that the total number of vertices of the 13 pieces of paper was 42.

How many triangles did she have before making the cuts?

A 8 B 7 C 6 D 5 E 4

SOLUTION **E**

Let s be the number of squares, and t the number of triangles that Mary started with. The number of vertices was $4s + 3t$. Also, there were 10 pieces of paper, so

$$s + t = 10. \qquad [1]$$

When she cuts a square diagonally to create two triangles, she increases the number of vertices by 2 (from 4 to 6). Hence before she cut the three squares, she had $42 - 3 \times 2 = 36$ vertices. Therefore

$$4s + 3t = 36. \qquad [2]$$

Subtracting three times equation [1] from equation [2] gives $s = 6$. Hence $t = 4$.

7. The positive integers a, b, c, d satisfy the equation $ab = 2cd$.

Which of the following numbers could not be the value of the product $abcd$?

A 50 B 100 C 200 D 450 E 800

SOLUTION **B**

Since $ab = 2cd$, the product $abcd = 2cd \times cd = 2(cd)^2$, hence it must be twice a perfect square. This is true for all the options, except 100 since $100 = 2 \times 50$ but 50 is not a perfect square. [$50 = 2 \times 5^2$; $200 = 2 \times 10^2$; $450 = 2 \times 15^2$; and $800 = 2 \times 20^2$.]

8. The shortest path from Atown to Cetown runs through Betown. Two of the signposts that can be seen on this path are shown, but one of them is broken and a number missing.

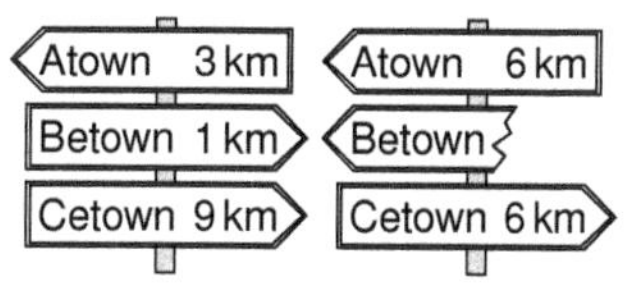

What distance was written on the broken sign?

A 2 km B 3 km C 4 km D 5 km E 6 km

SOLUTION 

The first signpost shows that Betown is 4 km from Atown. The second signpost is 6 km from Atown, so must be 2 km from Betown.

9. An isosceles triangle has a side of length 20 cm. Of the remaining two side-lengths, one is equal to two-fifths of the other. What is the length of the perimeter of this triangle?

A 36 cm B 48 cm C 60 cm D 90 cm E 120 cm

SOLUTION 

The triangle is isosceles so has a pair of equal sides. Since the two unknown sides are not equal, they cannot be the pair of equal sides, and hence the 20 cm side must be one of the pair of equal sides. The base is then two-fifths of 20 cm, namely 8 cm. The perimeter is $20 + 20 + 8 = 48$ cm.

10. Freda wants to write a number in each of the nine cells of this figure so that the sum of the three numbers on each diameter is 13 and the sum of the eight numbers on the circumference is 40.

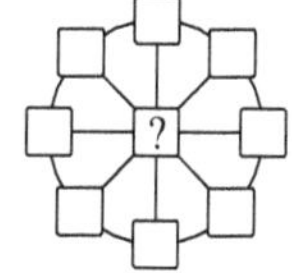

What number must be written in the central cell?

A 3 B 5 C 8 D 10 E 12

SOLUTION **A**

Each diameter has the same sum and contains the central cell, so the pair at the end of each diameter must have the same sum. These four pairs have sum 40, so each pair must have sum 10. Since each diameter has sum 13, the central number must be 3.

11. Masha put a multiplication sign between the second and third digits of the number 2020 and noted that the resulting product 20×20 was a square number.

How many integers between 2010 and 2099 (including 2020) have the same property?

A 1 B 2 C 3 D 4 E 5

SOLUTION **C**

Each number begins with 20, and $20 = 5 \times 2^2$. Hence, to make a square product, the last two digits must make a number which is a product of 5 and a square number. The possibilities between 10 and 99 are $5 \times 2^2 = 20$, $5 \times 3^2 = 45$ and $5 \times 4^2 = 80$. Therefore there are three possible numbers: 2020, 2045, 2080.

12. Two squares of different sizes are drawn inside an equilateral triangle. One side of one of these squares lies on one of the sides of the triangle as shown. What is the size of the angle marked by the question mark?

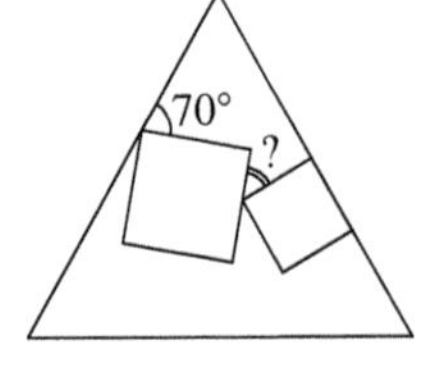

A 25° B 30° C 35° D 45° E 50°

SOLUTION **E**

On the diagram is a pentagon outlined, and four of its angles are known. The sum of the angles in a pentagon is 540°. The missing angle is then 540° −270° − 70° − 60° − 90° = 50°.

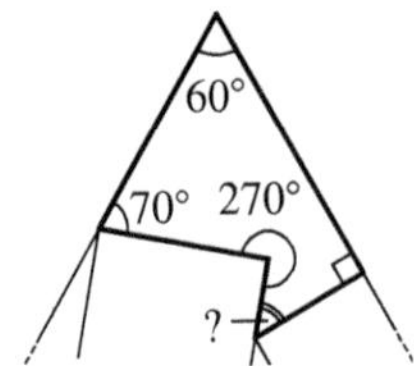

13. Luca began a 520 km trip by car with 14 litres of fuel in the car tank. His car consumes 1 litre of fuel per 10 km. After driving 55 km, he saw a road sign showing the distances from that point to five petrol stations ahead on the road. These distances are 35 km, 45 km, 55 km, 75 km and 95 km. The capacity of the car's fuel tank is 40 litres and Luca wants to stop just once to fill the tank.

How far is the petrol station that he should stop at?

A 35 km B 45 km C 55 km D 75 km E 95 km

SOLUTION **D**

Luca starts with 14 litres of fuel, which is enough for 140 km. After travelling 55 km, Luca can go a further 85 km. Hence, he cannot reach the 95 km petrol station, but can reach the others. If he stops at the 55 km petrol station (or any nearer one), then he will have at least 410 km left to travel of his 520 km journey, but his tank only holds enough for 400 km. Hence, he should stop at the 75 km petrol station, with 390 km left to travel.

14. The numbers x and y satisfy the equation $17x + 51y = 102$. What is the value of $9x + 27y$?

A 54 B 36 C 34 D 18
E The value is undetermined.

SOLUTION **A**

By dividing $17x + 51y = 102$ by 17 we get $x + 3y = 6$. Multiplying by 9 gives $9x + 27y = 54$.

15. A vertical stained glass square window of area 81 cm^2 is made out of six triangles of equal area (see figure). A fly is sitting on the exact spot where the six triangles meet. How far from the bottom of the window is the fly sitting?

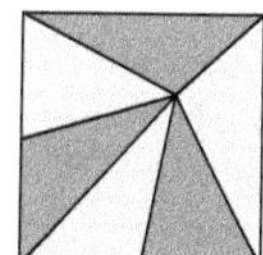

A 3 cm B 5 cm C 5.5 cm D 6 cm E 7.5 cm

SOLUTION **D**

Let h be the height of the fly above the base of the window. Each side-length of the square window of area 81 cm^2 is 9 cm. The two triangles that form the bottom part of the window have total area equal to a third of the whole window, namely 27 cm^2. Hence $\frac{1}{2} \times 9 \times h = 27$ so $h = 27 \times 2 \div 9 = 6$ cm.

16. The digits from 1 to 9 are randomly arranged to make a 9-digit number.

What is the probability that the resulting number is divisible by 18?

A $\frac{1}{3}$ B $\frac{4}{9}$ C $\frac{1}{2}$ D $\frac{5}{9}$ E $\frac{3}{4}$

SOLUTION **B**

To be divisible by 18, the number must be divisible by 2 and by 9. The digit sum for any number formed is $1 + 2 + 3 + 4 + 5 + 6 + 7 + 8 + 9 = 45$, and hence every number is a multiple of 9. Therefore any even number formed will be divisible by 18. Of the 9 possible last digits, there are four even digits (2, 4, 6, 8), so the probability of the number being even is $\frac{4}{9}$.

17. A hare and a tortoise competed in a 5 km race along a straight line, going due North. The hare is five times as fast as the tortoise. The hare mistakenly started running due East. After a while he realised his mistake, then turned and ran straight to the finish point. He arrived at the same time as the tortoise. What was the distance between the hare's turning point and the finish point?

A 11 km B 12 km C 13 km D 14 km E 15 km

SOLUTION **C**

The routes of the hare and the tortoise form a right-angled triangle as shown on the diagram. The hare travels 5 times as fast, but arrives at the same time as the tortoise, so has travelled five times further, giving

$$x + y = 25. \text{ [1]}$$

Pythagoras' Theorem gives $x^2 + 5^2 = y^2$, so $y^2 - x^2 = 25$, and this factorises to give $(y+x)(y-x) = 25$. However, $x+y = 25$ by equation [1]. So $25(y-x) = 25$ and then $y - x = 1$, and $y = x + 1$. Substituting this into equation [1] gives $x + x + 1 = 25$, and hence $x = 12$ and $y = 13$.

18. There are some squares and triangles on the table. Some of them are blue and the rest are red. Some of these shapes are large and the rest are small. We know that

1. If the shape is large, it's a square;
2. If the shape is blue, it's a triangle.

Which of the statements A–E must be true?

A All red figures are squares.
B All squares are large.
C All small figures are blue.
D All triangles are blue.
E All blue figures are small.

SOLUTION **E**

From statement 1 it follows that if a shape is not square, then it is not large; hence all triangles are small. Using statement 2 we see that blue figures are triangles and so are small. This shows that E is true. Suppose we had a set of shapes which consisted of one small blue triangle, one small red triangle, one small red square and one large red square. Then this satisfies statements 1 and 2 but it does not satisfy A, B, C or D. So they are not true in general.

19. Two identical rectangles with sides of length 3 cm and 9 cm are overlapping as in the diagram. What is the area of the overlap of the two rectangles?

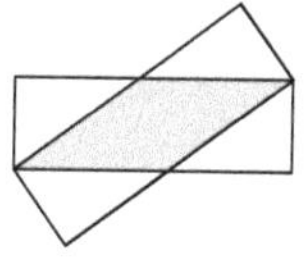

A 12 cm^2 B 13.5 cm^2 C 14 cm^2 D 15 cm^2
E 16 cm^2

SOLUTION **D**

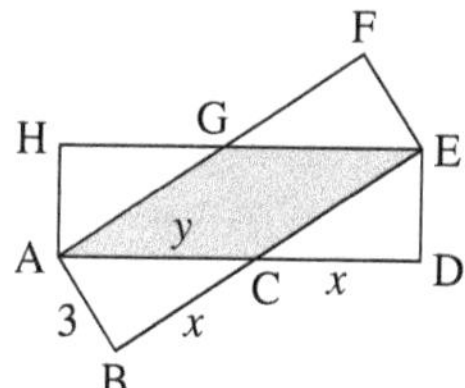

First we prove that the four white triangles are congruent. Note that they each have a right angle. Also angles HGA and FGE are equal (vertically opposite). The quadrilateral $ACEG$ is a parallelogram since each of its sides come from the rectangles. Hence angles FGE and GAC are equal (corresponding), and angles GAC and ECD are equal (corresponding). Therefore each triangle has a right-angle, and an angle equal to HGA, and therefore they each have the same angles. Moreover, they each have a corresponding side of length 3 cm. By labelling triangle ABC with lengths 3, x and y, Pythagoras' Theorem gives $x^2 + 3^2 = y^2$ [1]. The triangles ABC and CDE are congruent, so $CD = CB = x$. Since $AD = 9$, we can see $x + y = 9$ [2].

Equation [2] rearranges to $x = 9 - y$, so $x^2 = (9 - y)^2 = 81 - 18y + y^2$. Substituting this into [1] gives $81 - 18y + y^2 + 3^2 = y^2$, so $18y = 90$ and $y = 5$. Thus $x = 9 - 5 = 4$. The area of each white triangle is $\frac{1}{2} \times 3x = \frac{1}{2} \times 3 \times 4 = 6\,\text{cm}^2$. Thus the overlap is $3 \times 9 - 2 \times 6 = 15\,\text{cm}^2$.

20. Kanga labelled the vertices of a square-based pyramid using 1, 2, 3, 4 and 5 once each. For each face Kanga calculated the sum of the numbers on its vertices. Four of these sums equalled 7, 8, 9 and 10. What is the sum for the fifth face?

A 11 B 12 C 13 D 14 E 15

SOLUTION **C**

One face has total 7 which can only be obtained from the given numbers by adding 1, 2 and 4. Therefore it is a triangular face, and so the label of the top vertex, x say, is one of these three values. Hence the square face has 5 at one vertex. So the smallest possible face total for the square is $5 + 1 + 2 + 3 = 11$. Therefore the four face totals given, 7, 8, 9 and 10, must be the face totals of the triangular faces and their sum is 34. Note that each vertex except the top belongs to two triangles; and the top belongs to all four. So the sum 34 is twice the sum of all the labels plus an extra $2x$; that is $34 = 2(1 + 2 + 3 + 4 + 5) + 2x = 30 + 2x$. Hence $x = 2$ and the face total for the square face is $1 + 3 + 4 + 5 = 13$.

21. A large cube is built using 64 smaller identical cubes. Three of the faces of the large cube are painted. What is the maximum possible number of small cubes that can have exactly one face painted?

A 27 B 28 C 32 D 34 E 40

SOLUTION  **C**

The large cube is $4 \times 4 \times 4$. There are only two possible configurations of the three painted faces: either they all share a common vertex, or they don't.

When the three faces share a common vertex, the edges that they share will have more than one face painted, leaving 9 cubes on each of the three faces with exactly one face painted, as shown in the diagram on the left. This is 27 cubes altogether.

When the three faces don't share a common vertex, then each of the cubes on the common edges will have more than one face painted. This leaves two of the large faces having 12 cubes with one face painted, and the middle large face having 8 cubes, giving a total of 32 cubes, shown in the diagram on the right.

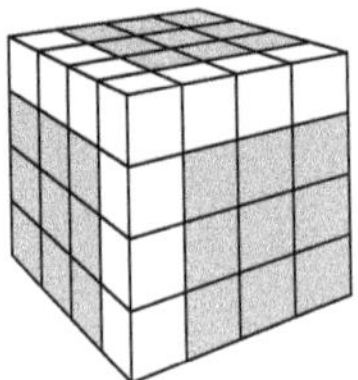

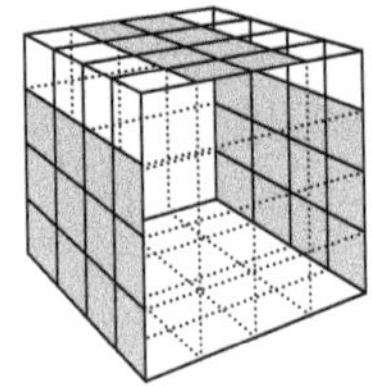

22. In each of the cells, a number is to be written so that the sum of the 4 numbers in each row and in each column are the same.

What number must be written in the shaded cell?

A 5 B 6 C 7 D 8 E 9

1		6	3
	2	2	8
	7		4
(shaded)		7	

SOLUTION 

Let x be the number in the bottom right cell. Then the column total is $x + 15$. Since each row and column has the same total, we can now find the other missing values. The third column requires x in its missing cell to make the total up to $x + 15$. The top row requires $x + 5$. The second column requires 1. The bottom row is now missing 7, hence this goes in the shaded square. [The other cells in the left column are $x + 3$ and 4.]

1	$x + 5$	6	3
$x + 3$	2	2	8
4	7	x	4
7 (shaded)	1	7	x

23. Alice, Belle and Cathy had an arm-wrestling contest. In each game two girls wrestled, while the third rested. After each game, the winner played the next game against the girl who had rested. In total, Alice played 10 times, Belle played 15 times and Cathy played 17 times. Who lost the second game?

A Alice
B Belle
C Cathy
D Either Alice or Belle could have lost the second game.
E Either Belle or Cathy could have lost the second game.

SOLUTION 

Since each game involved two girls, the number of games played is $(10 + 15 + 17) \div 2 = 21$. Alice played 10 of these, and rested for 11 of them. The maximum amount of resting possible is obtained by alternately losing and resting. To rest 11 times, Alice must have rested the odd-numbered games (1st, 3rd, etc) and lost the even numbered games (2nd, 4th, etc). Hence Alice lost the second game.

24. Eight consecutive three-digit positive integers have the following property: each of them is divisible by its last digit. What is the sum of the digits of the smallest of these eight integers?

A 9 B 10 C 11 D 12 E 13

SOLUTION **E**

Since we cannot divide by zero, the eight numbers must have the form ABn where n runs from 1 to 8 or from 2 to 9. However if ABn is divisible by n for each n from 2 to 9, it is also divisible by n for each n from 1 to 8 and $AB1$ is smaller than $AB2$. Hence our solution will use the digits 1 to 8. Note that ABn is divisible by n if and only if $AB0$ is divisible by n. So we require the smallest number $AB0$ which is divisible by 1, 2, 3, 4, 5, 6, 7, 8. The LCM of these 8 numbers is $8 \times 7 \times 5 \times 3 = 840$. The eight numbers are then 841, 842, 843, 844, 845, 846, 847, 848. The digit sum of 841 is 13.

25. A zig-zag line starts at the point P, at one end of the diameter PQ of a circle. Each of the angles between the zig-zag line and the diameter PQ is equal to α as shown. After four peaks, the zig-zag line ends at the point Q. What is the size of angle α?

not to scale

A 60° B 72° C 75° D 80° E 86°

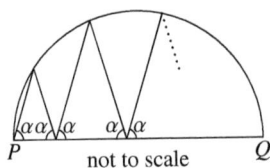

SOLUTION **B**

After four peaks the zig-zag is at the end of the diameter, so after two peaks it must be at the centre O of the circle. The triangle OPR is isosceles since OP and OR are both radii, hence angle $PRO = \alpha$ and angle $POR = 180° - 2\alpha$. Angle $OTR = 180° - \alpha$ (angles on a straight line), and hence angle $TRO = 3\alpha - 180°$ (angles in triangle TRO).

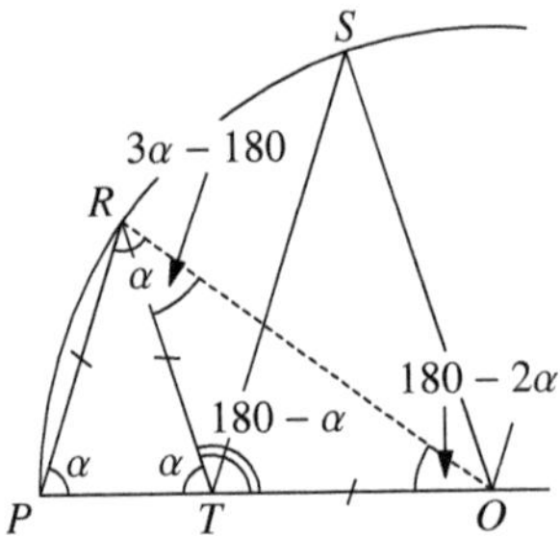

Triangle OTS is isosceles because its base angles are equal, and hence $ST = SO$. Therefore triangles OPR and OTS are congruent because they both have two sides which equal the radius of the circle, with angle $180° - 2\alpha$ between (SAS). Hence $OT = PR = TR$ (since $PR = RT$ in isosceles triangle PRT). Hence triangle OTR is isosceles, and its base angles are equal, that is $180° - 2\alpha = 3\alpha - 180°$. Solving this gives $\alpha = 72°$.

2.7.2 Participation

Participation in the Pink Kangaroo

Year of competition	2017	2018	2019	2020
Centres with invitees	1,737	1,834	1,803	1203
Centres returning scripts	1,470	1,519	1,603	996
Invitees	8,244	8,706	8,829	9165
Scripts returned	6,534	6,821	7,243	3890

Percentage change from previous year

Year of competition	2017	2018	2019	2020
Centres with invitees	+0.4%	+5.6%	-1.7%	-33.2 %
Centres returning scripts	+3.3%	+5.0%	+0.5%	-37.8 %
Invitees	+13.3%	+5.6%	+1.4%	+3.8%
Scripts returned	+21.9%	+4.4%	+6.2%	

2.7.3 Student performance

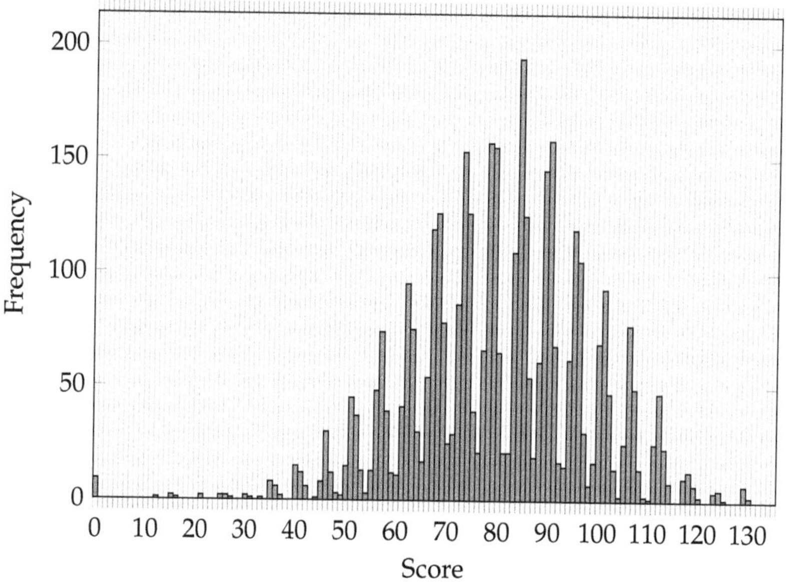

Answers given

The table below shows, for each question, the percentage of students giving each answer. The correct answer is underlined.

Question	A	B	C	D	E	Blank	Ambiguous
1	1	99	0	0	0	0	0
2	1	90	3	3	2	0	0
3	0	95	1	2	0	1	0
4	1	1	97	1	0	0	0
5	0	98	0	1	0	0	0
6	9	8	11	4	66	2	0
7	26	38	4	24	4	4	0
8	80	4	10	3	2	1	0
9	1	77	1	18	1	1	0
10	78	10	7	2	1	2	0
11	48	17	26	5	2	2	0
12	1	7	9	17	63	2	0
13	1	2	2	88	5	1	0
14	69	2	2	2	23	1	0
15	7	3	9	69	10	2	0
16	20	44	15	11	2	10	0
17	4	15	51	4	21	6	0
18	14	9	8	10	54	1	0
19	7	31	8	34	8	11	1
20	13	16	44	12	4	11	0
21	15	11	41	6	17	9	0
22	15	11	41	6	17	9	0
23	16	14	7	32	14	18	0
24	16	17	17	18	16	19	0
25	12	30	19	13	6	20	0

Mean scores and award boundaries

The table shows, for each of the past five years of the competition, the minimum score needed to obtain a Certificate of Merit and the mean score.

Year of competition	2016	2017	2018	2019	2020
Certificate of Merit	84	79	73	90	101
Mean score	71.9	67.0	60.6	77.1	80.0

2.8 Cayley Olympiad

2.8.1 Questions and solutions

United Kingdom
Mathematics Trust

INTERMEDIATE MATHEMATICAL OLYMPIAD

CAYLEY PAPER

Thursday 19 March 2020

supported by 

England & Wales: Year 9 or below
Scotland: S2 or below
Northern Ireland: Year 10 or below

These problems are meant to be challenging! The earlier questions tend to be easier; later questions tend to be more demanding.

Do not hurry, but spend time working carefully on one question before attempting another.

Try to finish whole questions even if you cannot do many: you will have done well if you hand in full solutions to two or more questions.

You may wish to work in rough first, then set out your final solution with clear explanations and proofs.

INSTRUCTIONS

1. Do not open the paper until the invigilator tells you to do so.
2. Time allowed: **2 hours**.
3. The use of blank or lined paper for rough working, rulers and compasses is allowed; **squared paper, calculators and protractors are forbidden**.
4. You should write your solutions neatly on A4 paper. Staple your sheets together in the top left corner with the Cover Sheet on top and the questions in order.
5. Start each question on a fresh A4 sheet. **Do not hand in rough work.**
6. Your answers should be fully simplified, and exact. They may contain symbols such as π, fractions, or square roots, if appropriate, but not decimal approximations.
7. You should give full written solutions, including mathematical reasons as to why your method is correct. Just stating an answer, even a correct one, will earn you very few marks; also, incomplete or poorly presented solutions will not receive full marks.

Enquiries about the Intermediate Mathematical Olympiad should be sent to:

UK Mathematics Trust, School of Mathematics, University of Leeds, Leeds LS2 9JT

☎ 0113 343 2339 enquiry@ukmt.org.uk www.ukmt.org.uk

1. In the triangle ABC, the three exterior angles $a°$, $b°$ and $c°$ satisfy $a + b = 3c$.

Prove that the triangle ABC is right-angled.

[Note: The diagram has been included to illustrate the labelling only and is not drawn to scale.]

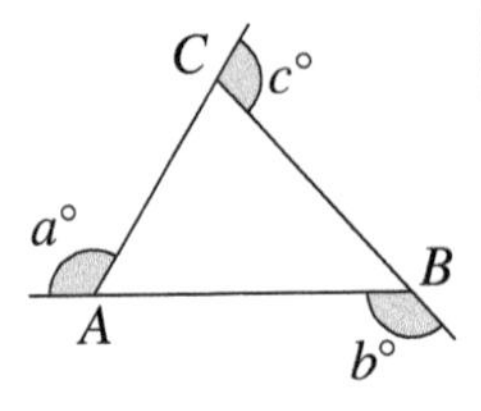

SOLUTION

The exterior and interior angles at each corner of the triangle add up to 180°, so the three interior angles have the values $180 - a$, $180 - b$ and $180 - c$.

The angles in any triangle add up to 180° so $(180 - a) + (180 - b) + (180 - c) = 180$.

Simplifying this equation gives $a + b + c = 360$.

The question tells us that $a + b = 3c$, so substituting for $a + b$ gives $3c + c = 360$, which simplifies to $4c = 360$.

Dividing both sides of this equation by 4 gives $c = 90$.

We know that the angle inside the triangle at C is equal to $180 - c = 180 - 90 = 90$. Therefore triangle ABC has a right angle at C.

2. The digits 1, 2, 3, 4, 5, A and B are all different and nonzero. Each of the two six-digit integers 'A12345' and '12345A' is divisible by B.

Find all possible pairs of values of A and B.

Solution

Since the digits 1, 2, 3, 4, 5, A and B are all different and nonzero, A and B must be two of 6, 7, 8 and 9.

The integer 'A12345' is odd and is divisible by B, so B must be odd too. This means B will either be 7 or 9.

Suppose $B = 7$. This means that '12345A' is divisible by 7, but the only multiples of 7 starting with 12345 are 123452 and 123459, which are 17636×7 and 17637×7, respectively. Since A cannot be 2, A must be 9.

We must now check that 'A12345' is divisible by 7 and this is true since 912345 is 130335×7.

Suppose instead that $B = 9$. This means that '12345A' is divisible by 9, but the only multiple of 9 starting with 12345 is 123453, which is 13717×9, and A cannot be 3, so B cannot be 9.

Therefore there is only one solution and that is $A = 9$, $B = 7$.

3. Four friends rent a cottage for a total of £300 for the weekend. The first friend pays half of the sum of the amounts paid by the other three friends. The second friend pays one third of the sum of the amounts paid by the other three friends. The third friend pays one quarter of the sum of the amounts paid by the other three friends.

How much money does the fourth friend pay?

Solution

Let f_a represent the amount of money the ath friend pays, in pounds. Therefore the sum of the amounts paid by the other three friends is $300 - f_a$ pounds.

The first friend pays half of the amount paid by the other three friends so $f_1 = \frac{1}{2}(300 - f_1)$. Multiplying both sides by 2 gives $2f_1 = 300 - f_1$. Adding f_1 to both sides gives $3f_1 = 300$ and so $f_1 = 100$.

Similarly, the second friend pays one third of the amount paid by the other three friends so $f_2 = \frac{1}{3}(300 - f_2)$. Multiplying both sides by 3 gives $3f_2 = 300 - f_2$. Adding f_2 to both sides gives $4f_2 = 300$ and so $f_2 = 75$.

Finally, the third friend pays a quarter of the amount paid by the other three friends so $f_3 = \frac{1}{4}(300 - f_3)$. Multiplying both sides by 4 gives $4f_3 = 300 - f_3$. Adding f_3 to both sides gives $5f_3 = 300$ and so $f_3 = 60$.

This means that the fourth friend must pay $300 - 100 - 75 - 60$ pounds.

So the fourth friend pays £65.

4. Two squares $ABCD$ and $BEFG$ share the vertex B, with E on the side BC and G on the side AB, as shown.
The length of CG is 9 cm and the area of the shaded region is 47 cm^2.

Calculate the perimeter of the shaded region.

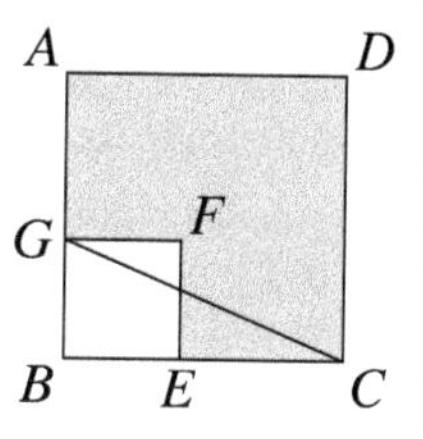

Solution

Let the length of BC be x cm.

Triangle BCG is right-angled and $CG = 9$ cm, so applying Pythagoras' Theorem we have $BG^2 = 9^2 - x^2 = 81 - x^2$.

The area of the shaded region = the area of square $ABCD$ – the area of square $BEFG$.

Therefore $47 = x^2 - (81 - x^2)$.

This simplifies to $47 = x^2 - 81 + x^2$.

Adding 81 to both sides and simplifying gives $128 = 2x^2$.

So $x^2 = 64$.

Since x represents a length, it cannot be negative, so $x = 8$.

Since $EF = BG$ and $FG = EB$, the perimeter of the shaded region is equal to the perimeter of square $ABCD$, which is 4×8.

Therefore the perimeter is 32 cm.

5. A ladybird is free to fly between the 1×1 cells of a 10×10 square grid. She may begin in any 1×1 cell of the grid. Every second she flies to a different 1×1 cell that she has not visited before.

Find the smallest number of cells the ladybird must visit, including her starting cell, so that you can be certain that there is a 2×2 grid of adjacent cells, each of which she has visited.

SOLUTION

Suppose each 1×1 cell starts coloured white and when the ladybird visits any cell, including the starting cell, a ladybird symbol is marked in that cell.

We will show that is is possible to have an arrangement of 75 ladybird symbols without a completely filled 2×2 grid anywhere but that as soon as you have any arrangement of 76 ladybird symbols, there must be a completely filled 2×2 grid somewhere in the arrangement.

It is possible to find many different arrangements of 75 ladybird symbols such that no 2×2 grid contains 4 ladybird symbols. For example:

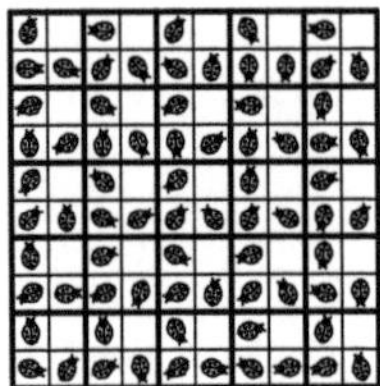
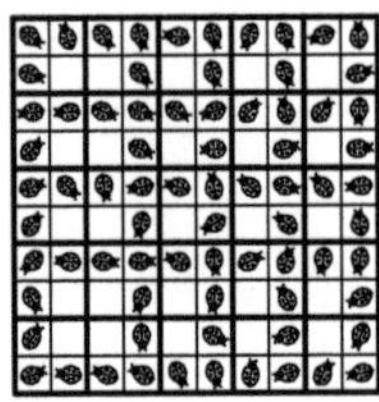

Since it is possible to use 75 ladybird symbols and not complete a 2×2 grid, the smallest number of cells visited that guarantees a completely filled 2×2 grid somewhere in the 10×10 grid must be greater than 75.

We now show that the smallest number to guarantee a complete 2×2 grid somewhere in the 10×10 grid is less than or equal to 76.

Split the 10×10 grid into 25 separate 2×2 grids.

The maximum number of symbols that these 25 separate 2×2 grids can hold without any of them being completely filled is 25×3, which is 75 ladybird symbols. So any attempt to insert 76 ladybird symbols into the 10×10 grid must completely fill one of these 2×2 grids.

Therefore, 76 is the smallest number of cells visited that guarantees a complete 2×2 grid somewhere.

6. Martha and Nadia play a game. Each has to make her own four-digit number, choosing her four digits from eight "digit cards" labelled 1-8. First Martha chooses her thousands digit, and then Nadia chooses her thousands digit. Next, Martha chooses her hundreds digit from the remaining six cards, and then Nadia chooses her hundreds digit. This process is repeated for the tens and finally the units digits of their numbers. The two four-digit numbers produced are then added together. Martha wins if the sum is not a multiple of 6; Nadia wins if the sum is a multiple of 6.

Determine which player has a winning strategy (that is to say, which player can guarantee that she will win no matter which digits the other player chooses).

Solution

For a number to be a multiple of 6, it must be a multiple of 2 and a multiple of 3.

Let Martha's four-digit number be '$abcd$' and Nadia's number be '$efgh$'.

We will show that the sum of these two four-digit numbers is actually always a multiple of 3.

The sum of the two four-digit numbers is $1000a + 100b + 10c + d + 1000e + 100f + 10g + h$.

This can be re-written as $999a + 99b + 9c + 999e + 99f + 9g + a + b + c + d + e + f + g + h$, which is $3(333a + 33b + 3c + 333e + 33f + 3g) + a + b + c + d + e + f + g + h$. This is a multiple of 3 added to $a + b + c + d + e + f + g + h$.

But we know that a, b, c, d, e, f, g and h are the digits 1, 2, 3, 4, 5, 6, 7 and 8 in some order. So $a+b+c+d+e+f+g+h = 1+2+3+4+5+6+7+8 = 36$, which is also a multiple of 3.

This means that the sum of the two four-digit numbers is always a multiple of 3, whatever cards Martha and Nadia choose.

So, for Nadia to win the game, she needs to be able to ensure that the sum of the two four-digit numbers is a multiple of 2, and so is automatically a multiple of 6.

A winning strategy for Nadia is to always choose an even number if Martha just chose an even number and an odd number if Martha chose an odd number. This is always possible as we start with four even and four odd numbers in the digits from 1 to 8. This strategy will guarantee that the sum of the two four-digit numbers is even and so a multiple of 6.

N.B. There are many winning strategies for Nadia, some of which replace the need to prove the general property that two four-digit numbers, chosen in the way described, always add up to a multiple of 3, with a restricted argument concerning particular pairings of numbers that tell Nadia what to choose if Martha chooses one of the digits in the pair. For example $\{1, 5\}$, $\{2, 8\}$, $\{3, 7\}$ and $\{4, 6\}$.

2.8.2 Participation

Participation in the Cayley Olympiad

Year of competition	2017	2018	2019	2020
Centres with invitees	270	264	258	146
Centres returning scripts	286	277	275	144
Invitees	521	544	527	481
Scripts returned	582	624	600	281

Percentage change from previous year

Year of competition	2017	2018	2019	2020
Centres with invitees	-2.2%	-2.2%	-2.3%	-43%
Centres returning scripts	+4.4%	-3.1%	-0.7%	-47.6%
Invitees	-1.5%	+4.4%	-3.1%	-8.7 %
Scripts returned	+0.0%	+7.2%	-3.8%	-47%

2.8.3 Student performance

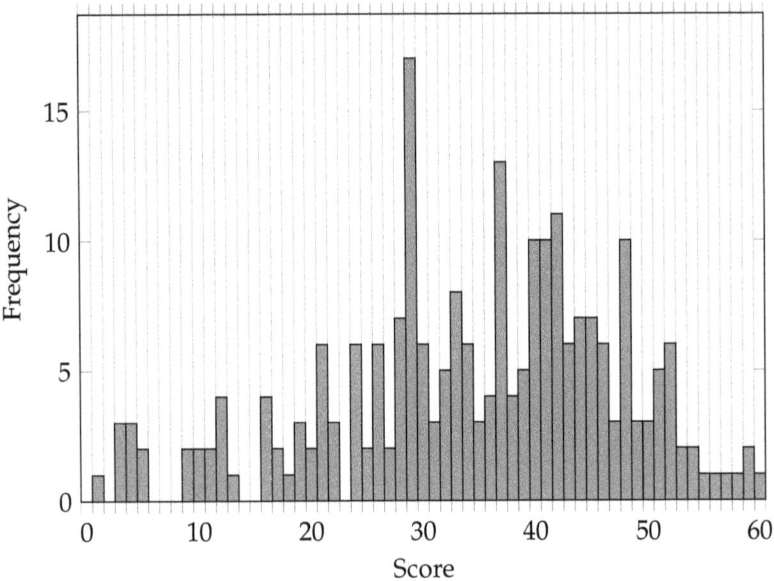

Mean scores and award boundaries

Each qualifying student receives a Certificate of Qualification and the highest-scoring participants receive Certificates of Distinction or Merit. The top 50 or so scorers are awarded a book prize; the title varies from year to year. We award a medal to the top 100 or so scorers in the competition.
The table below shows the minimum score needed to obtain each award, and the mean score, for each of the past five years of the competition.

Year of competition	2016	2017	2018	2019	2020
Prize	52	51	53	52	*
Medal	46	46	50	50	*
Certificate of Distinction	42	43	46	49	*
Certificate of Merit	10	10	10	10	*
Mean score	30.1	32.3	33.5	38.5	34.8

* Due to the COVID-19 pandemic, in 2020, prizes were not awarded.

Comments from the marking team

The unprecedented situation in the world at the time when this paper was taken did not stop those who sent in their scripts from producing many solutions of a high quality. The paper was found to be accessible by a large majority of the candidates, resulting in plenty of scripts containing five or six proposed solutions and many high scores overall. The markers enjoyed seeing clearly explained and well laid out solutions as well as unusual methods or observations. The inventiveness of some of the students remains refreshing and continues to be impressive. Questions 5 and 6 gave even the best students a chance to tackle challenging questions and there were some lovely solutions. Well done, students!

Question 1, *proposed by Gerry Leversha*
This question was relatively straight-forward, especially if students started by using the fact that the exterior angles of any polygon add to $360°$.

Question 2 *proposed by Dean Bunnell*
Once students understood properly what was being asked, solutions consisted of a mixture of helpful arguments that reduced the number of cases needing consideration (for example "Since B divides A12345, which is odd, B must itself be odd.") and checking particular cases. It should be noted that merely saying, for example, that a number is not divisible by 7, is not enough, markers would like to see a division sum, or a stated result to the division with a correct remainder. Also important to remember is that questions of this sort might have no solutions, so narrowing down the possibilities to one answer and then claiming that this must be the final answer is not enough, students must also show that the answer satisfies those conditions of the question not used in arriving at this possibility. For example, some students used the fact that 7 divides 912345 but needed to check that 7 also divides 123459 before claiming they were done.

Question 3 *proposed by Howard Groves*
This question was solved by many candidates, using a variety of algebraic routes, some considerably longer than others. Solving four equations in four unknowns is possible, but this was a much easier question for students who saw that the conditions about who paid what could be turned into single equations in one unknown each time. It would be good if students try to avoid saying "Let the first friend be a" when they mean "Let the amount paid by the first friend be a pounds" etc

Question 4 *proposed by Howard Groves*
This was a relatively easy Question 4 and many students correctly formed two equations – one by using Pythagoras, and the other by considering the shaded area to be the difference in the area of two squares. Solving the equations was easy but students should remember to say why they reject any negative solutions. Finally, some justification for the perimeter of the shaded area being the same as the perimeter of square ABCD was needed for full marks.

Question 5 *proposed by Tom Bowler*
This was a challenging question for many students. The most common error was to find a particular arrangement of 75 visited squares and then claim that visiting an-other square would definitely produce a completed

2x2 grid. But what if there was another arrangement of 75 visited squares that did allow a visit to another square without doing so? The argument for 76 visited squares meaning there must be a completed 2x2 grid must be a general one that covers any arrangement of 76 visited squares. Some students also tried to argue that every arrangement of 75 visited squares that had no completed 2x2 would have to have 3 out of 4 squares visited in every 2x2 grid. There are many arrangements where this is not the case, so these arguments had no foundation.

Question 6 *proposed by Tom Bowler*
The final question was completed well by a number of students. Unfortunately some students made the false claim that "the sum of the two four-digit numbers must be a multiple of 3 because 1 + 2 + 3 + ... + 8 = 36, which is a multiple of 3". They had not considered that there may be 'carries' from one column to the next on adding, which complicates the matter. In the end, all Nadia needs to do is ensure that the sum of the two four-digit numbers is even and this can be achieved in many different ways. It was good to see quite a few different valid winning strategies suggested.

Stephen Power
Cayley Olympiad Lead Marker

2.9 Hamilton Olympiad

2.9.1 Questions and solutions

United Kingdom
Mathematics Trust

INTERMEDIATE MATHEMATICAL OLYMPIAD

HAMILTON PAPER

supported by [XTX] MARKETS Overleaf

SOLUTIONS

These are polished solutions and do not illustrate the process of failed ideas and rough work by which candidates may arrive at their own solutions.

It is not intended that these solutions should be thought of as the 'best' possible solutions and the ideas of readers may be equally meritorious.

Enquiries about the Intermediate Mathematical Olympiad should be sent to:

UK Mathematics Trust, School of Mathematics, University of Leeds, Leeds LS2 9JT

☎ 0113 343 2339 enquiry@ukmt.org.uk www.ukmt.org.uk

Intermediate Mathematical Olympiad

Hamilton paper

Solutions

These are polished solutions and do not illustrate the process of failed ideas and rough work by which candidates may arrive at their own solutions.

It is not intended that these solutions should be thought of as the 'best' possible solutions and the ideas of readers may be equally meritorious.

Enquiries about the Intermediate Mathematical Olympiad should be sent to:

UK Mathematics Trust, School of Mathematics, University of Leeds, Leeds LS2 9JT

☎ 0113 343 2339 enquiry@ukmt.org.uk www.ukmt.org.uk

1. Arun and Disha have some numbered discs to share out between them. They want to end up with one pile each, not necessarily of the same size, where Arun's pile contains exactly one disc numbered with a multiple of 2 and Disha's pile contains exactly one disc numbered with a multiple of 3. For each case below, either count the number of ways of sharing the discs, or explain why it is impossible to share them in this way.

(a) They start with ten discs numbered from 1 to 10.

(b) They start with twenty discs numbered from 1 to 20.

SOLUTION

(a) Consider the disc numbered 6. Assume for the moment that it is in Arun's pile. Then all the other discs numbered with a multiple of 2 must be in Disha's pile.

Disha's pile must also contain one of the two remaining discs numbered with a multiple of 3 (with the other in Arun's pile), so there are ${}^2C_1 = 2$ ways to arrange the remaining discs numbered with a multiple of 3.

Once the discs containing multiples of 3 have been allocated to a pile, there are three remaining discs (those numbered 1, 5 and 7), each of which could go in either of the two piles, so there are $2^3 = 8$ ways this can be done.

So there are $2 \times 8 = 16$ ways of distributing the discs if the disc numbered 6 is in Arun's pile.

Now assume that the disc numbered 6 is in Disha's pile. Then all the other discs numbered with a multiple of 3 must be in Arun's pile.

Arun's pile must also contain one of the four remaining discs numbered with a multiple of 2 (with the other three all in Disha's pile), so there are ${}^4C_1 = 4$ ways to arrange the remaining discs numbered with a multiple of 2.

Once the discs containing multiples of 2 have been allocated to a pile, there are again three remaining discs (1, 5 and 7), and as before there are 8 ways these can be allocated to the piles.

So there are $4 \times 8 = 32$ ways of distributing the discs if the disc numbered 6 is in Disha's pile.

Hence there are $16 + 32 = 48$ ways they can share the discs out.

(b) Consider the three discs numbered 6, 12 and 18, which are multiples of both 2 and 3. Since there are three of them, when the discs are shared out, one of Arun's and Disha's pile must contain (at least) two of them. But that is against the rules for how Arun and Disha want to share the discs out, so it is impossible to share them out.

2. In the UK, 1p, 2p and 5p coins have thicknesses of 1.6 mm, 2.05 mm and 1.75 mm respectively.

Using only 1p and 5p coins, Joe builds the shortest (non-empty) stack he can whose height in millimetres is equal to its value in pence. Penny does the same but using only 2p and 5p coins.

Whose stack is more valuable?

SOLUTION

Say Joe has a 1p coins and b 5p coins. Then he wants the minimal a and b such that

$$a + 5b = 1.6a + 1.75b.$$

This simplifies to

$$12a = 65b.$$

Since a and b are integers, and 12 and 65 have no factors in common (other than 1), a must be a multiple of 65 and b must be a multiple of 12. Clearly the smallest (positive) a and b which satisfy this equation are $a = 65$ and $b = 12$. So Joe's stack comprises sixty-five 1p coins and twelve 5p coins, giving it a value of 125p.

Now say Penny has c 2p coins and d 5p coins. Then similarly, she wants the minimal c and d such that

$$2c + 5d = 2.05c + 1.75d,$$

which simplifies to

$$c = 65d.$$

The smallest positive c and d which satisfy this equation are $c = 65$ and $d = 1$. So Penny's stack comprises sixty-five 2p coins and one 5p coin, giving it a value of 135p.

Hence Penny's stack is more valuable.

3. The diagram shows two semicircles with a common centre O and a rectangle $OABC$. The line through O and C meets the small semicircle at D and the large semicircle at E. The lengths CD and CE are equal.

What fraction of the large semicircle is shaded?

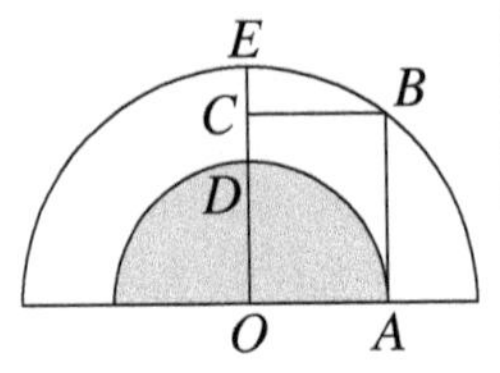

SOLUTION

Say $OA = r$ and $OB = R$. Then $OC = \frac{R+r}{2}$.

By Pythagoras' Theorem,

$$r^2 + (\frac{R+r}{2})^2 = R^2$$
$$3R^2 - 2Rr - 5r^2 = 0$$
$$(3R - 5r)(R + r) = 0$$

So (discounting $r = -R$, which is not practicable) $r = \frac{3}{5}R$.

Hence the fraction of the larger semicircle that is shaded is $(\frac{3}{5})^2 = \frac{9}{25}$.

4. Piercarlo chooses n integers from 1 to 1000 inclusive. None of his integers is prime, and no two of them share a factor greater than 1.

What is the greatest possible value of n?

SOLUTION

First, note that $31^2 = 961$ and $37^2 = 1369$, so the largest prime less than $\sqrt{1000}$ is 31.

So any non-prime number less than 1000 (excepting 1) must have at least one prime factor which is less than or equal to 31 (otherwise the number would be at least 37^2, which is larger than 1000).

We are told that all pairs of Piercarlo's share no factors greater than one; this clearly must include prime factors, and in particular the primes from 2 to 31 inclusive. Hence each of the primes up to 31 can only appear in the factorisation of one of his numbers.

So to maximise n, he will need to choose numbers with as few of these prime factors in their prime factorisation as possible. Indeed, he should aim to choose a set of numbers where each number he chooses has no more than one prime factor less than or equal to 31. Since there are 11 of these primes, the maximum value of n is 12 (since he can also pick 1).

It is easy to check that he can indeed pick 12 non-prime numbers less than 1000 – for example by picking 1 together with the squares of all the primes up to 31:

$$\{1, 4, 9, 25, 49, 121, 169, 289, 361, 529, 841, 961\}.$$

5. In the diagram, $ABCD$ is a parallelogram, M is the midpoint of AB and X is the point of intersection of AC and MD.

What is the ratio of the area of $MBCX$ to the area of $ABCD$?

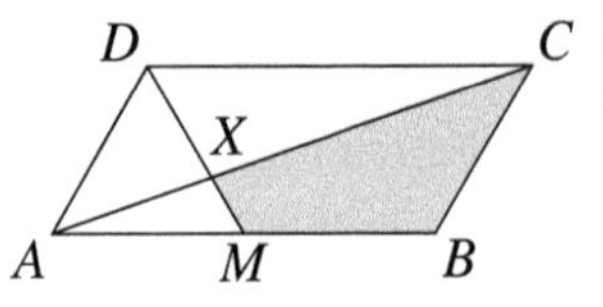

SOLUTION

Solution 1

Say that the area of triangle XAM is 1 square unit. Note that $\angle XMA = \angle XDC$ and $\angle XAM = \angle XCD$ (since they are alternate angles between parallel lines), so triangles XCD and XAM are similar. Since $DC = 2AM$, the area of XCD is $1 \times 2^2 = 4$ square units.

Say the area of triangle DAX is x square units. Then, since the area of ACD is twice the area of AMD (they are triangles with equal heights and one base half that of the other), we have $2 \times (1 + x) = 4 + x$, which gives $x = 2$ square units.

The area of $ABCD$ is twice that of ACD, which is 12 square units.

The area of $MBCX$ is $12 - (4 + 2 + 1) = 5$ square units.

So the desired ratio is 5:12.

Solution 2

Note that $\angle XMA = \angle XDC$ and $\angle XAM = \angle XCD$ (since they are alternate angles between parallel lines), so triangles XCD and XAM are similar. Since $DC = 2AM$, the perpendicular from DC to X is twice the perpendicular from AM to X.

Hence the perpendicular height of the triangle AMX is one third of the perpendicular height of the parallelogram. Hence the area of AMX is $\frac{1}{2} \times \frac{1}{2} \times \frac{1}{3} = \frac{1}{12}$ of the area of the parallelogram.

The area of triangle ABC is clearly half of the area of the parallelogram, so the area of $MBCX$ is $\frac{1}{2} - \frac{1}{12} = \frac{5}{12}$ of the area of the whole parallelogram.

Hence the desired ratio is 5:12.

6. We write $\lfloor x \rfloor$ to represent the largest integer less than or equal to x. So, for example, $\lfloor 1.7 \rfloor = 1$, $\lfloor 2 \rfloor = 2$, $\lfloor \pi \rfloor = 3$ and $\lfloor -0.4 \rfloor = -1$.

Find all real values of x such that $\lfloor 3x + 4 \rfloor = \lfloor 5x - 1 \rfloor$.

SOLUTION

Solution 1

First note that $x - 1 < \lfloor x \rfloor \le x$ (*).

Also note that, for any integer n, $\lfloor x + n \rfloor = \lfloor x \rfloor + n$.

So the equation $\lfloor 3x + 4 \rfloor = \lfloor 5x - 1 \rfloor$ can be rearranged to give

$$\lfloor 5x \rfloor - \lfloor 3x \rfloor = 5.$$

Now, by (*), $5x - (3x - 1) > 5$, which simplifies to $x > 2$, and $(5x - 1) - 3x < 5$, which simplifies to $x < 3$.

Given $2 < x < 3$, the value of $\lfloor 3x \rfloor$ will only change as x changes value above/below $2\frac{1}{3}$ and $2\frac{2}{3}$.

Similarly, the value of $\lfloor 5x \rfloor$ will only change as x changes value above/below $2\frac{1}{5}$, $2\frac{2}{5}$, $2\frac{3}{5}$ and $2\frac{4}{5}$.

So now we just need to consider the regions between 2 and 3 created by these values:

x	$\lfloor 3x + 4 \rfloor$	$\lfloor 5x - 1 \rfloor$
$2 < x < 2\frac{1}{5}$	10	9
$2\frac{1}{5} \le x < 2\frac{1}{3}$	10	10
$2\frac{1}{3} \le x < 2\frac{2}{5}$	11	10
$2\frac{2}{5} \le x < 2\frac{3}{5}$	11	11
$2\frac{3}{5} \le x < 2\frac{2}{3}$	11	12
$2\frac{2}{3} \le x < 2\frac{4}{5}$	12	12
$2\frac{4}{5} \le x < 3$	12	13

So the set of values for which the equation holds is {x such that $2\frac{1}{5} \le x < 2\frac{1}{3}$ or $2\frac{2}{5} \le x < 2\frac{3}{5}$ or $2\frac{2}{3} \le x < 2\frac{4}{5}$}.

Solution 2

First note that $x - 1 < \lfloor x \rfloor \leq x$ (*).

Say that $\lfloor 3x + 4 \rfloor = k$ for some integer k. Then from (*) we have $3x + 4 \geq k$ and $3x + 3 < k$. These can be rearranged and combined to give

$$\frac{k-4}{3} \leq x < \frac{k-3}{3}.$$

Now we also have $\lfloor 5x - 1 \rfloor = k$, so again from (*) we have $5x - 2 < k$ and $5x - 1 \geq k$, which can be rearranged and combined to give

$$\frac{k+1}{5} \leq x < \frac{k+2}{5}.$$

The solutions to the equation will be where these two regions of the number line overlap.

First, we need the two regions on the number line to overlap at all, so both:

$$\frac{k+1}{5} < \frac{k-3}{3}, \text{ which gives } k > 9,$$

$$\text{and } \frac{k-4}{3} < \frac{k+2}{5}, \text{ which gives } k < 13.$$

We can check the three remaining values of k:

When $k = 10$, the overlap is $2\frac{1}{5} \leq x < 2\frac{1}{3}$.

When $k = 11$, the overlap is $2\frac{2}{5} \leq x < 2\frac{3}{5}$.

And when $k = 12$, the overlap is $2\frac{2}{3} \leq x < 2\frac{4}{5}$.

These three regions of overlap are exactly the regions where the equation is satisfied.

2.9.2 Participation

Participation in the Hamilton Olympiad

Year of competition	2017	2018	2019	2020
Centres with invitees	286	286	261	159
Centres returning scripts	291	300	278	166
Invitees	549	510	497	477
Scripts returned	650	594	595	326

Percentage change from previous year

Year of competition	2017	2018	2019	2020
Centres with invitees	+9.6%	+0.0%	-8.7%	-39%
Centres returning scripts	+11.1%	+3.1%	-7.3%	-40.2%
Invitees	+4.6%	-7.1%	-2.5%	-0.4%
Scripts returned	+14.6%	-8.6%	+0.2%	-55%

2.9.3 Student performance

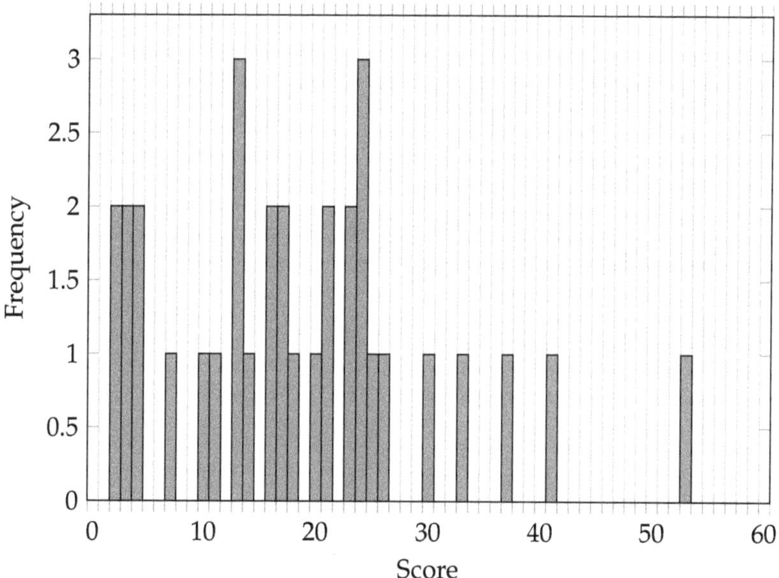

Mean scores and award boundaries

Each qualifying student receives a Certificate of Qualification and the highest-scoring participants receive Certificates of Distinction or Merit. The top 50 or so scorers are awarded a book prize; the title varies from year to year. We award a medal to the top 100 or so scorers in the competition.
The table below shows the minimum score needed to obtain each award, and the mean score, for each of the past five years of the competition.

Year of competition	2017	2018	2019	2020
Prize	53	52	38	*
Medal	48	46	31	*
Certificate of Distinction	41	41	28	*
Certificate of Merit	10	10	8	*
Mean score	29.8	29.4	19.7	18.6

* Due to the COVID-19 pandemic, prizes were not awarded in 2020.

Comments from the marking team

Given the highly unusual circumstances, it was very pleasing to see how many candidates had still been able to have a go at some (hopefully!) interesting and challenging problems. The variety of solutions and innovativeness shown within them was genuinely inspiring.

Question 1 *proposed by Daniel Griller*
Being the first question on the paper, this was designed to be accessible. Pleasingly, almost every candidate attempted this question and there were lots of very fine solutions produced. Pretty much all solutions focused on the disc numbered 6, but there were other successful methods employed. Most candidates also explained their methods clearly and thoroughly. The most common error was deciding that discs 1, 5 and 7 could be shared out in six different ways (presumably because this is 3x2) as opposed to eight (being 2^3).

Question 2 *proposed by Daniel Griller*
Again, this question was attempted by the vast majority of candidates and there were many excellent solutions submitted, almost all of which

formed a pair of Diophantine equations (one for each pile) and solved them (although some did work with the difference between a coin's value and its height). There were many candidates who didn't justify why their solutions gave the smallest such towers of coins.
Several candidates used x, y and z (or similar) to represent the number of 1p, 2p and 5p coins, but used z in both equations (i.e. used it to represent two different unknowns), which is to be discouraged. There was also a number of students whose solutions left lots of gaps for the reader to fill in! The best work ensured all unknowns were carefully defined and all equations justified. A minority argued that each value of coins had to produce an integer-height stack, which is not the case – even those who reached the right answer by this method did not earn that many marks because their justification was flawed.

Question 3 *proposed by Will Taylor*
All approaches seen here were essentially the same, labelling two different lengths (often the two radii, as in the official solutions, but sometimes DC or DE) then using Pythagoras' Theorem in triangle OAB to reach a quadratic equation.
It was very pleasing how many candidates were undeterred by this equation's unusual appearance and solved it successfully, although a small number did divide through by (e.g.) $(R - r)$ without explicitly checking that this was non-zero.
Some candidates recognised that, since this is a question about ratios, we can fix one length (e.g. OA) to be 1, however, some scripts didn't justify this choice, which did cost a couple of marks.

Question 4 *proposed by Tom Bowler*
I can imagine there are a lot of candidates here who feel hard done by, having scored only two or three out of ten. The difficulty that almost all candidates had was separating the two aspects required in a solution. The first of these was constructing a set which worked for Piercarlo to maximise n (many did this very well). The second was justifying how we know that there cannot possibly be a larger one, and it was this that was found to be challenging.
Most candidates dove straight into a construction (invariably using the squares of the primes up to 31), giving a vague attempt at an explanation along the lines of "we have to choose numbers with the fewest prime

factors" (which is not the case – consider the set 1, 841, 893, 901, 949, 961, 973, 979, 989, 993, 995, 998, nine of which have two prime factors but which is still a set of twelve numbers Piercarlo could have chosen).
The key idea, which some solutions did manage to explain satisfactorily, is that none of Piercarlo's numbers can share any prime factor less than or equal to 31. Once this is established, an upper bound of $n = 12$ is quickly reached, at which point constructing a set of 12 numbers is relatively straightforward.

Question 5 *proposed by Gerry Leversha*
A lot more candidates had success in Q5 than Q4. Almost all argued using similar triangles $XAM \sim XCD$, although it was important to explain how it was known that they were similar (some solutions made this claim without any justification at all). It is also important good practice to indicate the angle fact being used (e.g. $\angle CAM = \angle ACD$ by the theorem of alternate angles).
As in Q3, it was acceptable to fix one length (since the question is interested in a ratio), provided doing this was justified properly. However, a handful of candidates went too far doing this, with some fixing AD and DC to be in one particular ratio (e.g. by forcing D to be directly above M). At least one candidate claimed that we could assume $ABCD$ was a rectangle (which, granted, gets the answer, but the approach only works if one infers from the question that the answer exists, which really isn't in the spirit of the paper).

Question 6 *proposed by Tom Bowler*
This question was, by design, very challenging, but that did not dissuade a decent chunk of candidates from giving it a go. It was almost a shame it came at the end because there were several attempts which showed promise but were clearly produced under time pressure and consequently didn't make enough progress to earn many marks.
The two approaches given in the official solutions were both seen in candidates' scripts.
Some candidates found the correct set of values for x but didn't convincingly justify how they knew it was complete (several just stopped below 2 and above 3 with no reason given at all). Some worked "outwards" from 2.5 (the solution to the equation without the floor function), but those that did rarely managed to explain successfully why

they stopped looking at some point. Some attempted the problem graphically, which was an excellent way to grasp what's going on but proved more difficult to explain formally and completely.

James Hall
Hamilton Olympiad Lead Marker

2.10 Maclaurin Olympiad

2.10.1 Question paper and solutions

United Kingdom
Mathematics Trust

INTERMEDIATE MATHEMATICAL OLYMPIAD

MACLAURIN PAPER

Thursday 19 March 2020

supported by [XTX] MARKETS Overleaf

England & Wales: Year 11
Scotland: S4
Northern Ireland: Year 12

These problems are meant to be challenging! The earlier questions tend to be easier; later questions tend to be more demanding.

Do not hurry, but spend time working carefully on one question before attempting another.

Try to finish whole questions even if you cannot do many: you will have done well if you hand in full solutions to two or more questions.

You may wish to work in rough first, then set out your final solution with clear explanations and proofs.

INSTRUCTIONS

1. Do not open the paper until the invigilator tells you to do so.
2. Time allowed: **2 hours**.
3. The use of blank or lined paper for rough working, rulers and compasses is allowed; **squared paper, calculators and protractors are forbidden**.
4. You should write your solutions neatly on A4 paper. Staple your sheets together in the top left corner with the Cover Sheet on top and the questions in order.
5. Start each question on a fresh A4 sheet. **Do not hand in rough work**.
6. Your answers should be fully simplified, and exact. They may contain symbols such as π, fractions, or square roots, if appropriate, but not decimal approximations.
7. You should give full written solutions, including mathematical reasons as to why your method is correct. Just stating an answer, even a correct one, will earn you very few marks; also, incomplete or poorly presented solutions will not receive full marks.

Enquiries about the Intermediate Mathematical Olympiad should be sent to:

UK Mathematics Trust, School of Mathematics, University of Leeds, Leeds LS2 9JT

☎ 0113 343 2339 enquiry@ukmt.org.uk www.ukmt.org.uk

Intermediate Mathematical Olympiad

Maclaurin paper

Solutions

These are polished solutions and do not illustrate the process of failed ideas and rough work by which candidates may arrive at their own solutions.

It is not intended that these solutions should be thought of as the 'best' possible solutions and the ideas of readers may be equally meritorious.

Enquiries about the Intermediate Mathematical Olympiad should be sent to:

UK Mathematics Trust, School of Mathematics, University of Leeds, Leeds LS2 9JT

☎ 0113 343 2339 enquiry@ukmt.org.uk www.ukmt.org.uk

1. A bag contains counters, of which ten are coloured blue and Y are coloured yellow. Two yellow counters and some more blue counters are then added to the bag. The proportion of yellow counters in the bag remains unchanged before and after the additional counters are placed into the bag.

Find all possible values of Y.

Solution

Let X be the number of blue counters added to the bag. The proportion of yellow counters in the bag at the beginning is

$$\frac{Y}{Y+10}$$

and the proportion of yellow counters in the bag at the end is

$$\frac{Y+2}{X+Y+12}$$

Hence we have

$$\frac{Y}{10+Y} = \frac{Y+2}{X+Y+12}$$

Cross-multiplying, we obtain $Y(X+Y+12) = (Y+2)(10+Y)$, which gives

$$XY + Y^2 + 12Y = 10Y + Y^2 + 20 + 2Y$$

This simplifies to $XY = 20$. Hence the number of yellow counters originally in the bag is either $1, 2, 4, 5, 10$ or 20.

(Two other valid equations, $\frac{Y}{10} = \frac{Y+2}{X+10}$ and $\frac{Y}{10} = \frac{2}{X}$, also lead to $XY = 20$)

2. In the square $ABCD$, the bisector of $\angle CAD$ meets CD at P and the bisector of $\angle ABD$ meets AC at Q.

What is the ratio of the area of triangle ACP to the area of triangle BQA?

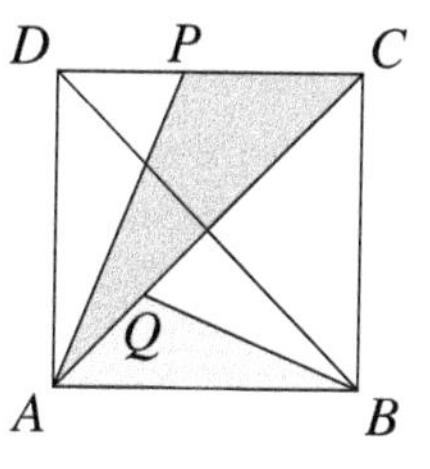

SOLUTION

We know that, since AC is the diagonal of a square, $\angle PCA = 45° = \angle QAB$. Also $\angle DAC = 45° = \angle DBA$. Since AP and BQ are angle bisectors, $\angle PAC = 22.5° = \angle QBA$.

Hence the triangles CPA and AQB have equal angles, so they are similar. Now $CA = \sqrt{2}$ and $AB = 1$, so the sides are in the ratio $\sqrt{2} : 1$, and the areas are in the ratio $2 : 1$.

3. An altitude of a triangle is the shortest distance from a vertex to the line containing the opposite side.

Find the side lengths of all possible right-angled triangles with perimeter 5 cm and shortest altitude 1 cm.

SOLUTION

The three altitudes of a right-angled triangle are the two sides around the right angle and the perpendicular from the right angle to the hypotenuse, which is clearly the shortest.

Let the lengths of the three sides in cm be a, b, c, where c is the hypotenuse.

(a) From the perimeter, we know that $a + b + c = 5$.

(b) The area using base a and height b is $\frac{1}{2}ab$ and the area using base c and height 1 is $\frac{1}{2}c$, so we obtain $ab = c$.

(c) From Pythagoras, we have $a^2 + b^2 = c^2$.

From (a) we have $a + b = 5 - c$, which, when squared, gives $a^2 + 2ab + b^2 = 25 - 10c + c^2$.

Substituting from (b) and (c), this becomes $c^2 + 2c = 25 - 10c + c^2$, so $12c = 25$ and $c = \frac{25}{12}$.

Now we know that $a + b = \dfrac{35}{12}$ and $ab = \dfrac{25}{12}$.

Hence $a + \dfrac{25}{12a} = \dfrac{35}{12}$ and so $12a^2 - 35a + 25 = 0$, which has solutions $a = \frac{5}{3}$ or $\frac{5}{4}$.

Finally the corresponding values of b are $\frac{5}{4}$ and $\frac{5}{3}$. Hence there is a unique triangle with sides $\frac{5}{3}$ cm, $\frac{5}{4}$ cm and $\frac{25}{12}$ cm.

An alternative approach uses $a + b + ab = 5$ to obtain $a = \dfrac{5 - b}{1 + b}$.

Hence $(\frac{5-b}{1+b})^2(b^2 - 1) = b^2$, so we have $(5 - b)^2(b - 1) = b^2(1 + b)$ and this, when expanded, gives $12b^2 - 35b + 25 = 0$ and so $(3b - 5)(4b - 5) = 0$.

Hence $b = \frac{5}{3}$ or $\frac{5}{4}$ and we find a and c as before.

Since this argument is not obviously reversible, it is necessary to check that these lengths satisfy the requirements of the problem.

4. The diagram shows a triangle ABC and two lines AD and BE, where D is the midpoint of BC and E lies on CA. The lines AD and BE meet at Z, the midpoint of AD.

What is the ratio of the length CE to the length EA?

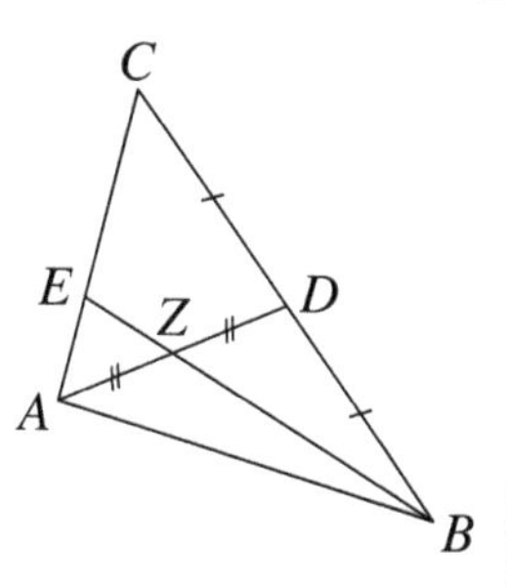

SOLUTION

A **Using areas**

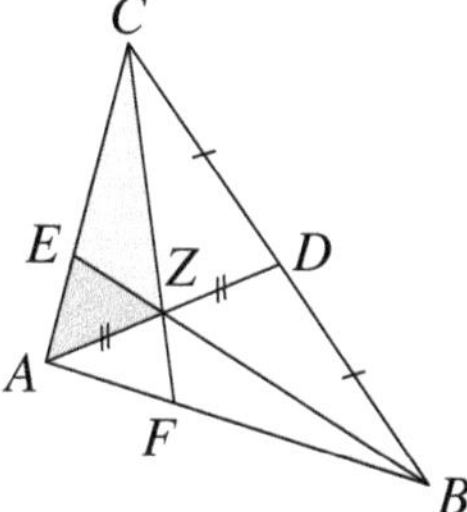

Since $CD = DB$, $[CDZ] = [BDZ]$ and, since $DZ = ZA$, $[BDZ] = [BAZ]$.

Now, considering triangles on the side CA, we have $\dfrac{[CEZ]}{[AEZ]} = \dfrac{[CZB]}{[AZB]} = 2$, and it follows that $CE : EA = 2 : 1$.

What is being used here is the useful 'arrowhead' shortcut which links the shaded areas directly to the unshaded ones. It is more likely that solvers will proceed via the step $\dfrac{[CEZ]}{[AEZ]} = \dfrac{[CEB]}{[AEB]}$ and then use a bit of algebra. Alternatively, some might be aware that if $\dfrac{a}{b} = \dfrac{c}{d}$, it follows that $\dfrac{a}{b} = \dfrac{a-c}{b-d}$.

B **Using vectors**

Let A be the origin and denote B and C by $4\mathbf{b}$ and $4\mathbf{c}$ respectively. Then D is $2(\mathbf{b} + \mathbf{c})$ and Z is $\mathbf{b} + \mathbf{c}$. Now E lies on both BZ and AC, so there are real numbers λ, μ such that $4\mu\mathbf{c} = \lambda(\mathbf{b} + \mathbf{c}) + 4(1 - \lambda)\mathbf{b}$. Hence $\lambda = \frac{4}{3}$ and $\mu = \frac{1}{3}$, so $CE : AE = 2 : 1$.

C **Using construction**

Let X be on BA such that $BA = AX$. Then CA is a median of triangle BCX.

The triangles BDA and BCX are similar, by ratios, so BE, which is a median of BDA, is also a median of BCX.

Hence E is the centroid of BCX and so $CE : EA = 2 : 1$

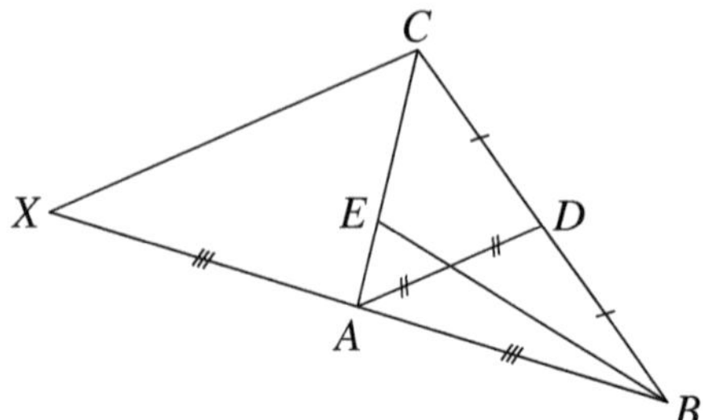

There are several other constructions which work. Other approaches use coordinate geometry, vectors and complex numbers. Ratio theorems such as Ceva and Menelaus are also useful.

5. Let p and q respectively be the smallest and largest prime factors of n. Find all positive integers n such that $p^2 + q^2 = n + 9$.

SOLUTION

First eliminate the special case where $n = p = q$, which leads to $2p^2 = p + 9$. This has no integer solutions.

Rewrite the equation as $p^2 + q^2 - 9 = kpq$ for some positive integer k

Now, since $q \mid q^2$ and $q \mid kpq$, we have $q \mid p^2 - 9 = (p-3)(p+3)$ and, since q is prime, $q \mid p-3$ or $q \mid p+3$

Hence either $p = 3$ or $q \mid p + 3$

If $p = 3$ then $n = q^2$ and $3 \mid n$ so $q = 3$ and $n = 9$, which works.

If $q \mid p + 3$ and $p = 2$ then $q = 5$ and $n = 20$, which works.

If $p > 3$ then $q > \frac{p+3}{2}$ and so $q = p + 3$, which is even. So this is impossible.

Hence only solutions are $n = 9$ and $n = 20$.

An alternative approach, using modular arithmetic, observes that $p^2 \equiv 9 \pmod{q}$

Since q is prime, this implies that $p \equiv \pm 3 \pmod{q}$ and then proceeds as above.

6. Seth and Cain play a game. They take turns, and on each turn a player chooses a pair of integers from 1 to 50. One integer in the pair must be twice the other, and the players cannot choose any integers used previously. The first player who is unable to choose such a pair loses the game. If Seth starts, determine which player, if any, has a winning strategy.

SOLUTION

Seth will win if the total number of moves is odd, and Cain if it is even.

A player must choose k and $2k$ for some $1 \leq k \leq 25$. This leads to the partition of the integers from 1 to 50 shown below. At any turn, a player must choose consecutive numbers from the same set.

$$\{1\ 2\ 4\ 8\ 16\ 32\} \quad \{5\ 10\ 20\ 40\}$$

$$\{3\ 6\ 12\ 24\ 48\}$$

$$\{7\ 14\ 28\} \quad \{9\ 18\ 36\} \quad \{11\ 22\ 44\}$$

$$\{13\ 26\} \quad \{15\ 30\} \quad \{17\ 34\} \quad \{19\ 38\} \quad \{21\ 42\} \quad \{23\ 46\} \quad \{25\ 50\}$$

The remaining twelve integers are in singleton sets and they cannot be chosen in the game.

The 10 sets consisting of two or three elements can each only be chosen once. As their number is even, this makes no difference to the result.

The five-element set containing 3 can only be used twice, so this too is irrelevant.

So if the players stick to these 11 sets, they will occupy 12 moves.

Hence the outcome of the game rests entirely on how the players use the sets. We call moves from these two sets 'interesting'.

Seth can guarantee a win if, on his first move, he chooses either $(1, 2)$ or $(16, 32)$ from the first interesting set. The two sets which remain after this removal have exactly the same structure, namely $\{k, 2k, 4k, 8k\}$. As soon as Cain chooses a pair from one of them, Seth mimics him by choosing the corresponding pair from the other. As a result, these two sets are exhausted in an odd number of moves, so the total number of moves is odd and Seth wins.

Note that it is not essential that Seth chooses $(1, 2)$ or $(16, 32)$ immediately, but it is harder to analyse the alternatives. If he is the first to choose a pair from the interesting sets, he will lose if it is anything but $(1, 2)$ or $(16, 32)$. If Cain is the first to choose such a pair, he will lose if it is anything but $(1, 2)$ or $(16, 32)$. If, however, he chooses one of these pairs, Seth can still win by being patient. He can temporise with uninteresting moves until Cain is forced to play a second interesting move once all the uninteresting moves have been used. Then Seth is back in control and can force a win.

2.10.2 Participation

Participation in the Maclaurin Olympiad

Year of competition	2017	2018	2019	2020
Centres with invitees	265	282	275	154
Centres returning scripts	278	303	296	186
Invitees	504	518	486	442
Scripts returned	600	608	574	340

Percentage change from previous year

Year of competition	2017	2018	2019	2020
Centres with invitees	+6.4%	-2.5%	-2.5%	-44%
Centres returning scripts	-1.1%	+9.0%	-2.3%	-33.7%
Invitees	-4.7%	+2.8%	-6.2%	-0.9%
Scripts returned	-2.6%	+1.3%	-5.6%	-60%

2.10.3 Student performance

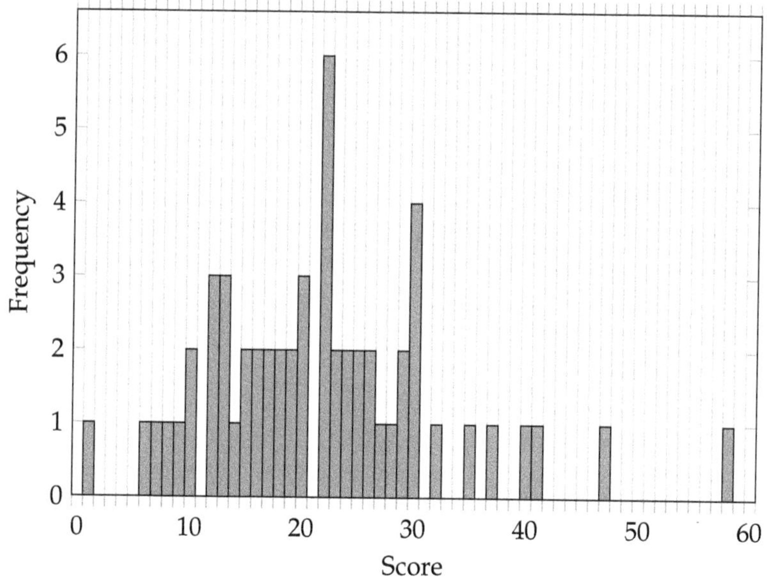

Mean scores and award boundaries

The table below shows, for each of the past five years of the competition, the minimum score needed to obtain each award and the mean score.

Year of competition	2017	2018	2019	2020
Prize	41	35	43	*
Medal	32	28	38	*
Certificate of Distinction	26	24	33	*
Certificate of Merit	9	8	10	*
Mean score	18.7	18.0	23.9	22.2

* Due to the COVID-19 pandemic, prizes were not awarded in 2020.

Comments from the marking team

Question 1 *proposed by Michael Griffiths*
This question was answered confidently by many candidates. The information given is that two ratios are equal. It is important to understand that a numerical ratio such as $a : b$ can be represented (in more than one way) as a fraction – for example, as $\frac{a}{b}$ or as $\frac{a}{a+b}$.
The fact that two ratios $a : b$ and $c : d$ are equal is now expressible as an algebraic equation $\frac{a}{b} = \frac{c}{d}$ which can be manipulated according to generic rules to obtain $ad = bc$. The facts given in this problem allowed for a variety of approaches, all of which eventually led to the fact that the product of two positive integers is 20.
Some candidates tried to work directly with ratios, attempting to make $Y + Y + 10$ and $Y + 2 : Y + X + 12$ look 'the same'. This led, essentially, to trial and improvement methods, which were successful in obtaining a number of solutions, but failed to demonstrate that all solutions to the problem had been found.

Question 2 *proposed by Gerry Leversha*
This question is about areas of triangles. It is tempting, in such a context, to make calculations, using formulae such as $A = \frac{1}{2}bh$ or $A = \frac{1}{2}ab \sin c$.
This approach worked well but candidates needed to be careful about some of the assumptions they made, particularly concerning the length of the segments PC or QB.
Some candidates used the angle bisector theorem to obtain these lengths correctly. In fact, hardly any calculation is needed, since it is enough to

evaluate two easy angles (45° and 22.5°) in triangles APC and BQA. It follows that these triangles are similar. The ratio of sides is $\sqrt{2}:1$ and therefore the ratio of areas is the square of this, namely 2:1 . The majority of successful solutions did this competently.
A principal source of error was to assume that the line which bisects an angle of a triangle also bisects the opposite side. Unfortunately this approach was taken by several unsuccessful candidates (leading to the correct ratio of 2:1 but achieving no marks). There were a few solvers who claimed that the triangles were congruent and they were docked a mark. It is also worth emphasising that the order of vertices is important when two triangles are described as similar. In this case, APC and BQA are similar but APC and BAQ are not, since P in the first triangle does not correspond to A in the other. Candidates who did this were gently reminded of the convention but were not penalised in the marking.

Question 3 *proposed by Tom Bowler*
This question was found to be challenging. It requires the solver to derive three equations from the information given (($a+b+c=5$) from the perimeter $a^2+b^2=c^2$, from Pythagoras $ab=c$, and , which is discussed later) and process them algebraically. This is quite a lot of work and you need to be confident and careful. There is quite a neat solution which begins by showing that c is $\frac{25}{12}$ and then calculating a and b . A number of successful candidates managed to derive a cubic expression for b and solve it.
Unfortunately the majority of candidates did not get as far as the algebra, since they did not understand the concept of a 'shortest altitude'. With hindsight, perhaps it was misleading to set a problem in which the word 'shortest' occurs twice. An altitude is the shortest distance from a vertex to the opposite side. This requires you to think about 'dropping a perpendicular' from a vertex, and, as there are three vertices, there are three altitudes. In a right-angled triangle, this is a little confusing, since two of the altitudes are also sides, but the third altitude, from the right angle to the hypotenuse, is not. However, it is this altitude which is the shortest of the three, since it is clearly shorter than the two sides around the right angle.
There were plenty of scripts which assumed that a (the shortest side of the triangle) was the altitude in question and hence obtained $b+c=4$ as an equation. These scripts were given a compensatory mark if they carried the algebra through from there. However, the correct way of

treating this fact is to calculate the area of a right-angled triangle in two ways: as $\frac{1}{2}ab$ and as $\frac{1}{2}ch$.
In the absence of an areal argument, several candidates looked at the similar triangles formed by the altitude. If it divides the hypotenuse into segments of lengths p and q, then it is easy to see $pq = 1$ that but much harder to solve the equation in these variables for the perimeter of the triangle. One candidate did manage to construct a cubic, two of whose roots were p and q, which allowed the proof to be completed. It is also possible, with ingenuity, to use p and q to show that $ch = ab$.
Perhaps more time ought to be devoted in the classroom to this concept of 'height'. There are nice exercises which depend on it, notably finding an expression for the inradius of a triangle, or asking which route a man marooned in a field bordered by three straight roads should take to hitch a lift home from a passing motorist. This sort of thing might not be in the syllabus, but it is certainly worth exploring with an able class.

Question 4 *proposed by Alan Slomson*
This question was, on the contrary, much better answered than we had anticipated. There were a variety of approaches, including areas, vectors, and constructions with a great deal of ingenuity being displayed. The only remark from markers is they would really appreciate being shown the diagram which is clearly in the candidate's rough working but not on display in the script.

Question 5 *proposed by Tom Bowler*
This is a technical problem and only a handful of students scored full marks. It is unfortunate that the case $n = p = q$ had not been explicitly ruled out in the statement of the question; it is clearly impossible but it needs to be addressed. The 'elegant' solution focuses on q and exploits the fact that the 'difference of two squares' is lurking in the statement of the problem: since $q|n$, we know that $q|(p-3)(p+3)$. Hence we have two cases, and, since we know that $p < q$, these become $p = 3$ and $q|p+3$. The first of these produces the solution $n = 9$ and the second, aided by considerations of magnitude, gives $n = 20$.
If you don't spot this, then you are in for quite a lot of argument by cases, which depend on the value of p and its relationship to q . The easiest to deal with are $p = 2$ and $p = 3$ which rapidly produce the two solutions for n . However, the hardest case to process is when $q > p > 3$ where the

argument is delicate, and unfortunately this meant that few candidates achieved the 10- threshold. There are also some overlaps in the assumptions for partial solutions. Two of these are the cases $q = p > 2$ and $q > p > 2$, the second of which is close to the 'elegant' solution.

Question 6 *proposed by Tom Bowler*
This problem can be approached without prior knowledge of game theory.
From Seth's point of view, he aims to be the last to play a move, and, as he has the first move, he wants the total number of moves to be odd. Cain, on the other hand, requires the opposite to be true. This, of course, depends on what numbers are still available at various stages of the game, and what choices are made, and so this is the first thing to investigate.
At the beginning there are 50 numbers, but nobody can choose any of the twelve odd numbers greater than 25. Then there are seven pairs, beginning 13, 26 and ending 25, 50 which can only be played once. There are three triples of numbers, such as 7, 14, 28, which can only be utilised once, even though there is a choice of which pair to play. There is one set of five numbers, namely 3 6 12 24 48, which will always yield two plays. All of these sets of numbers taken together will result in precisely twelve moves.
There remain two sets, namely 1 2 4 8 16 32 and 5 10 20 40 where there is an element of choice. The first will yield either two or three moves, and the second either one or two moves. Seth wants to control these two sets to produce an odd number of moves.
Now we have the moment of insight. If Seth removes either the first pair or the last pair from the set 1 2 4 8 16 32, this will leave two quadruples of numbers with exactly the same structure. He can now mimic whatever Cain does, and as a result he will have the final move. So his winning strategy is to act in this way, and he would be wise to do so immediately.
The threshold set for this question was to show the partition of the 50 numbers into sets and to identify the two sets which are critical in determining Seth's strategy. Many candidates had the partition. What the marker had to establish was whether they realised that everything boiled down to the two key sets. If a candidate made it clear that these sets were significant, they would score 7 marks even if they then wrongly advised Seth to play 2 4.
Finishing the question was, of course, another matter. If anyone mentioned the element of 'strategy-stealing' then this was an indication

that the candidate knew exactly what was going on. In some cases, it was not stated that Seth should play his crucial move immediately. In such cases, we could not assume that Cain would be a gentleman and wait for him to do so, and we then needed an explanation of how Seth could still win. It turns out that it depends on whether Cain does what Seth ought to have done or does something else. At least one script omitted a set from the partition, as an oversight. This would, of course, mean a victory for Cain, but that did not affect the fact that the candidate had achieved the threshold.

This was quite a daring question to set, but it evoked a good response and, with luck, some fruitful discussion afterwards.

Gerry Leversha
Maclaurin Olympiad Lead Marker

2.11 Senior Mathematical Challenge

2.11.1 Question paper and solutions

United Kingdom
Mathematics Trust

SENIOR MATHEMATICAL CHALLENGE

Thursday 7 November 2019

Organised by the United Kingdom Mathematics Trust

supported by  

Candidates must be full-time students at secondary school or FE college.
England & Wales: Year 13 or below
Scotland: S6 or below
Northern Ireland: Year 14 or below

INSTRUCTIONS

1. Do not open the paper until the invigilator tells you to do so.
2. Time allowed: **90 minutes**.
   No answers, or personal details, may be entered after the allowed time is over.
3. The use of blank or lined paper for rough working is allowed; **squared paper**, **calculators and measuring instruments are forbidden**.
4. **Use a B or an HB non-propelling pencil.** Mark at most one of the options A, B, C, D, E on the Answer Sheet for each question. Do not mark more than one option.
5. **Do not expect to finish the whole paper in the time allowed.** The questions in this paper have been arranged in approximate order of difficulty with the harder questions towards the end. You are not expected to complete all the questions during the time. You should bear this in mind when deciding which questions to tackle.
6. **Scoring rules:**
   All candidates start with 25 marks;
   0 marks are awarded for each question left unanswered;
   4 marks are awarded for each correct answer;
   1 mark is deducted for each incorrect answer (to discourage guessing).
7. Your Answer Sheet will be read by a machine. **Do not write or doodle on the sheet except to mark your chosen options.** The machine will read all black pencil markings even if they are in the wrong places. If you mark the sheet in the wrong place, or leave bits of eraser stuck to the page, the machine will interpret the mark in its own way.
8. **The questions on this paper are designed to challenge you to think, not to guess.** You will gain more marks, and more satisfaction, by doing one question carefully than by guessing lots of answers. This paper is about solving interesting problems, not about lucky guessing.

Enquiries about the Senior Mathematical Challenge should be sent to:

UK Mathematics Trust, School of Mathematics, University of Leeds, Leeds LS2 9JT

☎ 0113 343 2339 enquiry@ukmt.org.uk www.ukmt.org.uk

1. What is the value of $123^2 - 23^2$?

A 10 000 B 10 409 C 12 323 D 14 600 E 15 658

2. What is the value of $(2019 - (2000 - (10 - 9))) - (2000 - (10 - (9 - 2019)))$?

A 4040 B 40 C −400 D −4002 E −4020

3. Used in measuring the width of a wire, one mil is equal to one thousandth of an inch. An inch is about 2.5 cm.

Which of these is approximately equal to one mil?

A $\frac{1}{40}$ mm B $\frac{1}{25}$ mm C $\frac{1}{4}$ mm D 25 mm E 40 mm

4. For how many positive integer values of n is $n^2 + 2n$ prime?

A 0 B 1 C 2 D 3 E more than 3

5. Olive Green wishes to colour all the circles in the diagram so that, for each circle, there is exactly one circle of the same colour joined to it.

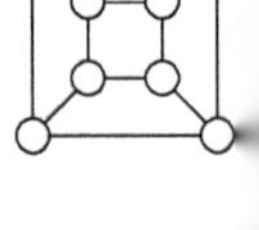

What is the smallest number of colours that Olive needs to complete this task?

A 1 B 2 C 3 D 4 E 5

6. Each of the factors of 100 is to be placed in a 3 by 3 grid, one per cell, in such a way that the products of the three numbers in each row, column and diagonal are all equal. The positions of the numbers 1, 2, 50 and x are shown in the diagram.

x	1	50
2		

What is the value of x?

A 4 B 5 C 10 D 20 E 25

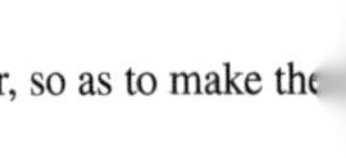

7. Lucy is asked to choose p, q, r and s to be the numbers 1, 2, 3 and 4, in some order, so as to make the value of $\frac{p}{q} + \frac{r}{s}$ as small as possible.

What is the smallest value Lucy can achieve in this way?

A $\frac{7}{12}$ B $\frac{2}{3}$ C $\frac{3}{4}$ D $\frac{5}{6}$ E $\frac{11}{12}$

8. The number x is the solution to the equation $3^{(3^x)} = 333$.

Which of the following is true?

A $0 < x < 1$ B $1 < x < 2$ C $2 < x < 3$ D $3 < x < 4$ E $4 < x < 5$

9. A square of paper is folded in half four times to obtain a smaller square. Then a corner is removed as shown.

Which of the following could be the paper after it is unfolded?

A 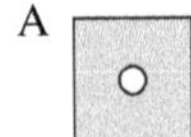 B 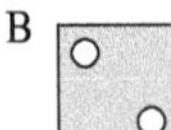 C 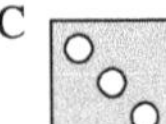 D  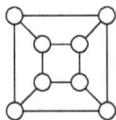 E 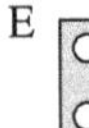

10. Which of the following five values of n is a counterexample to the statement in the box below?

For a positive integer n, at least one of $6n - 1$ and $6n + 1$ is prime.

A 10 B 19 C 20 D 21 E 30

11. For how many integer values of k is $\sqrt{200 - \sqrt{k}}$ also an integer?

A 11 B 13 C 15 D 17 E 20

12. A circle with radius 1 touches the sides of a rhombus, as shown. Each of the smaller angles between the sides of the rhombus is 60°.

What is the area of the rhombus?

A 6 B 4 C $2\sqrt{3}$ D $3\sqrt{3}$ E $\frac{8\sqrt{3}}{3}$

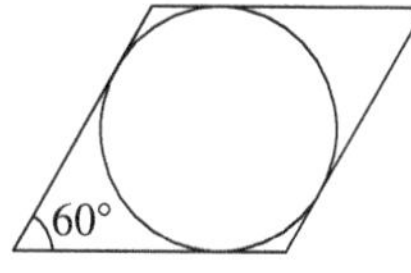

13. Anish has a number of small congruent square tiles to use in a mosaic. When he forms the tiles into a square of side n, he has 64 tiles left over. When he tries to form the tiles into a square of side $n + 1$, he has 25 too few.

How many tiles does Anish have?

A 89 B 1935 C 1980 D 2000 E 2019

14. One of the following is the largest square that is a factor of 10!. Which one?

Note that, $n! = 1 \times 2 \times 3 \times \cdots \times (n - 1) \times n$.

A $(4!)^2$ B $(5!)^2$ C $(6!)^2$ D $(7!)^2$ E $(8!)^2$

15. The highest common factors of all the pairs chosen from the positive integers Q, R and S are three different primes.

What is the smallest possible value of $Q + R + S$?

A 41 B 31 C 30 D 21 E 10

16. The numbers x, y and z satisfy the equations $9x + 3y - 5z = -4$ and $5x + 2y - 2z = 13$.

What is the mean of x, y and z?

A 10 B 11 C 12 D 13 E 14

17. Jeroen writes a list of 2019 consecutive integers. The sum of his integers is 2019.

What is the product of all the integers in Jeroen's list?

A 2019^2 B $\frac{2019 \times 2020}{2}$ C 2^{2019} D 2019 E 0

18. Alison folds a square piece of paper in half along the dashed line shown in the diagram. After opening the paper out again, she then folds one of the corners onto the dashed line.

What is the value of α?

A 45 B 60 C 65 D 70 E 75

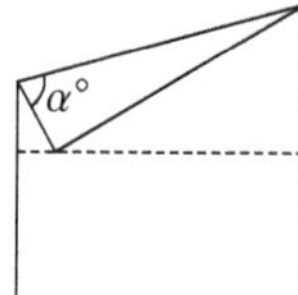

19. Which of the following could be the graph of $y^2 = \sin(x^2)$?

A

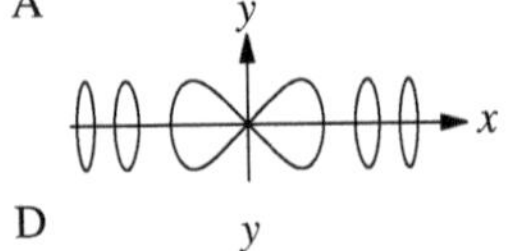

B

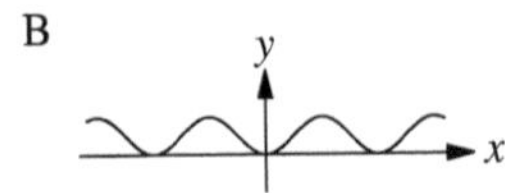

C

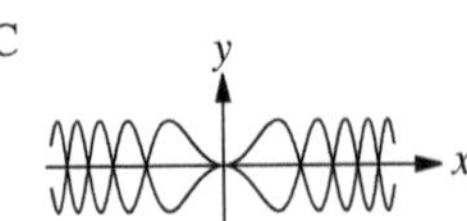

D

E

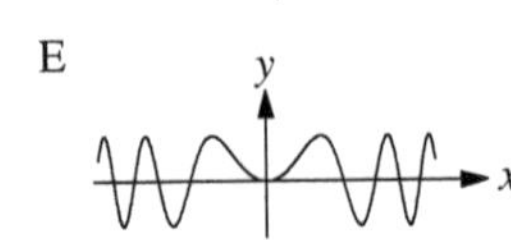

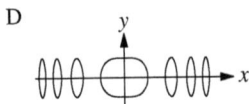

20. The "heart" shown in the diagram is formed from an equilateral triangle ABC and two congruent semicircles on AB. The two semicircles meet at the point P. The point O is the centre of one of the semicircles. On the semicircle with centre O, lies a point X. The lines XO and XP are extended to meet AC at Y and Z respectively. The lines XY and XZ are of equal length.

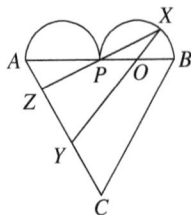

What is $\angle ZXY$?

A $20°$ B $25°$ C $30°$ D $40°$ E $45°$

21. In a square garden $PQRT$ of side 10 m, a ladybird sets off from Q and moves along edge QR at 30 cm per minute. At the same time, a spider sets off from R and moves along edge RT at 40 cm per minute.

What will be the shortest distance between them, in metres?

A 5 B 6 C $5\sqrt{2}$ D 8 E 10

22. A function f satisfies the equation $(n - 2019)f(n) - f(2019 - n) = 2019$ for every integer n.

What is the value of $f(2019)$?

A 0 B 1 C 2018×2019 D 2019^2 E 2019×2020

23. The edge-length of the solid cube shown is 2. A single plane cut goes through the points Y, T, V and W which are midpoints of the edges of the cube, as shown.

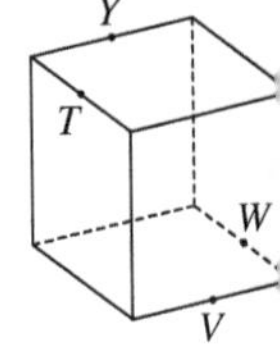

What is the area of the cross-section?

A $\sqrt{3}$ B $3\sqrt{3}$ C 6 D $6\sqrt{2}$ E 8

24. The numbers x, y and z are given by $x = \sqrt{12 - 3\sqrt{7}} - \sqrt{12 + 3\sqrt{7}}$, $y = \sqrt{7 - 4\sqrt{3}} - \sqrt{7 + 4\sqrt{3}}$ and $z = \sqrt{2 + \sqrt{3}} - \sqrt{2 - \sqrt{3}}$.

What is the value of xyz ?

A 1 B -6 C -8 D 18 E 12

25. Two circles of radius 1 are such that the centre of each circle lies on the other circle. A square is inscribed in the space between the circles.

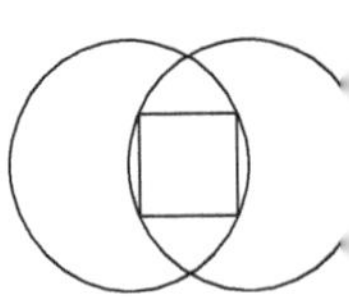

What is the area of the square?

A $2 - \frac{\sqrt{7}}{2}$ B $2 + \frac{\sqrt{7}}{2}$ C $4 - \sqrt{5}$ D 1 E $\frac{\sqrt{5}}{5}$

Senior Mathematical Challenge

Thursday 7 November 2019

For reasons of space, these solutions are necessarily brief.

There are more in-depth, extended solutions available on the UKMT website, which include some exercises for further investigation:

www.ukmt.org.uk

1. D The value of $123^2 - 23^2 = (123 - 23)(123 + 23) = 100 \times 146 = 14600$.

2. B The value of $(2019-(2000-(10-9)))-(2000-(10-(9-2019))) = (2019-1999)-(2000-2020) = 20 - (-20)$ which equals 40.

3. A One mil $= \frac{1}{1000}$ in $\approx \frac{1}{1000} \times 2.5\,\text{cm} = \frac{25}{1000}\,\text{mm} = \frac{1}{40}\,\text{mm}$.

4. B The expression $n^2 + 2n$ factorises to $n(n + 2)$. For $n(n + 2)$ to be prime, one factor must equal 1 whilst the other must be equal to a prime. This happens when $n = 1$, as $n + 2 = 3$, but not when $n + 2 = 1$ as n would be negative. There is therefore exactly one positive integer value of n which makes $n^2 + 2n$ prime.

5. B Each circle in the diagram is connected to three others, exactly one of which must be filled with the same colour. So, the number of colours required is greater than 1. One possible colouring with just two colours is shown here.

6. D The product of all the factors of 100 is $1 \times 100 \times 2 \times 50 \times 4 \times 25 \times 5 \times 20 \times 10 =$ 1 000 000 000. As there are three rows, each of which has the same 'row product', that row product is 1000. So, considering the top row, $x \times 1 \times 50 = 1000$ and therefore $x = 20$. The completed grid is as shown.

20	1	50
25	10	4
2	100	5

7. D In order to minimise the value of $\frac{p}{q} + \frac{r}{s}$ we need to make p and r as small as possible and make q and s be as large as possible. Considering $\frac{1}{3} + \frac{2}{4}$ and $\frac{1}{4} + \frac{2}{3}$ and then removing both $\frac{1}{3}$ and $\frac{1}{4}$ from each sum leaves the first with value $\frac{1}{4}$ and the second with value $\frac{1}{3}$. As $\frac{1}{4} < \frac{1}{3}$, the first sum has the smallest value, which is $\frac{5}{6}$.

8. B Considering some integer powers of 3, we have $3^5 = 243$ and $3^6 = 729$. As $243 < 333 < 729$, $3^5 < 3^{(3^x)} < 3^6$ implies that $5 < 3^x < 6$. Rewriting once again to include powers of 3, gives $3^1 < 5 < 3^x < 6 < 3^2$, so $3^1 < 3^x < 3^2$ and finally, $1 < x < 2$.

9. D Folding the paper four times gives 2^4 layers. Removing a corner, 16 quarter-circles are formed. Of the given options, only D, with four whole circles could then be possible.

10. C To provide a counterexample, we are looking for both the values of $6n - 1$ and $6n + 1$ to be composite for a particular n. When $n = 20$, $6n - 1 = 119 = 7 \times 17$ and $6n + 1 = 121 = 11 \times 11$, which is our counter-example. In each of the other cases, at least one value is prime.

11. C In order for $\sqrt{200-\sqrt{k}}$ to be an integer, $200-\sqrt{k}$ must be a square. As $\sqrt{k} \geq 0$, the smallest possible value of $200-\sqrt{k}$ which is square is 0, (when $k = 200^2$) and the largest is $14^2 = 196$, (when $k = 4^2$). Counting the squares from 0^2 to 14^2 gives 15 values.

12. E Let the centre of the rhombus and circle be O. Let two of the vertices along an edge of the rhombus be P and Q and let X be the point on PQ where the rhombus is tangent to the circle. In order to relate the radius of the inscribed circle to a useful measurement on the rhombus, we can split the rhombus along its diagonals into four congruent triangles, one of which is POQ. As PQ is tangent to the circle at X, PQ and OX are perpendicular. Triangles OXP and OXQ are then similar 30°, 60°, 90° triangles with $OX = 1$, $XP = \sqrt{3}$ and $XQ = \frac{1}{\sqrt{3}}$. The area of the rhombus is then $4 \times \frac{1}{2} \times (\sqrt{3} + \frac{1}{\sqrt{3}}) \times 1 = 2 \times \frac{(3+1)}{\sqrt{3}} = \frac{8}{3}\sqrt{3}$.

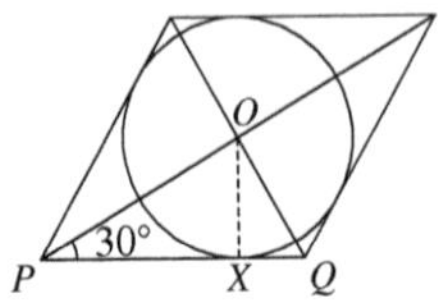

13. D Anish has both $n^2 + 64$ and $(n+1)^2 - 25$ tiles. So, $n^2 + 64 = (n+1)^2 - 25$ which simplifies to $n^2 + 64 = n^2 + 2n + 1 - 25$ and then $64 = 2n - 24$. So $n = \frac{88}{2} = 44$. Therefore Anish has $44^2 + 64 = 1936 + 64 = 2000$ tiles.

14. C For a square to be a factor of 10!, the prime factors of the square must be present in 10! an even number of times. Writing 10! first as $10 \times 9 \times 8 \times 7 \times 6 \times 5 \times 4 \times 3 \times 2 \times 1$ and then as $2 \times 5 \times 3 \times 3 \times 2 \times 2 \times 2 \times 7 \times 2 \times 3 \times 5 \times 2 \times 2 \times 3 \times 2 = 2^8 \times 3^4 \times 5^2 \times 7 = (2^4 \times 3^2 \times 5)^2 \times 7$ we can see that the largest square is $2^4 \times 3^2 \times 5$. Expanding $2^4 \times 3^2 \times 5$ as $(2 \times 3) \times 5 \times (2 \times 2) \times 3 \times 2 \times 1$ shows that it is exactly 6!. So $(6!)^2$ is the largest square which is a factor of 10!.

15. B For $Q + R + S$ to be as small as possible, we want the highest common factors of the pairs to be as small as possible, and prime. Therefore the highest common factors are 2, 3 and 5 in some order and then Q, R and S are 2×3, 2×5 and 3×5, i.e. 6, 10 and 15, in some order. This gives $Q + R + S = 6 + 15 + 10 = 31$.

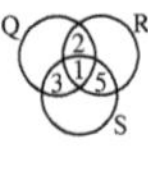

16. A As $9x + 3y - 5z = -4$ and $5x + 2y - 2z = 13$, subtracting the first equation from twice the second gives $(10x + 4y - 4z) - (9x + 3y - 5z) = 2 \times 13 - (-4)$. The mean of x, y and z which is $\frac{(x+y+z)}{3}$ is therefore $\frac{30}{3} = 10$.

17. E The sum of the first 2019 positive integers is $\frac{2019 \times 2020}{2}$ which is considerably larger than the required sum of 2019. In order for the sum of Jeroen's 2019 integers to be only 2019, some of the integers must be positive and some must be negative. One of the integers will then be 0, so the product will also be 0. Jeroen's list is $-1008, \ldots -2, -1, 0, 1, 2, \ldots, 1008, 1009, 1010$.

18. E Let P and Q be the vertices on the top edge of the square. Let R and S be the points at the end of the first fold line. Let P' be the position of P on the first fold line after the second fold has been made. Let T be the point on PS which lies on the second fold line. Triangles PQT and $P'QT$ are then congruent so $\angle PQT = \angle P'QT = 90° - \alpha°$. As $PQ = 2QR$ then $P'Q = 2QR$ and so $\angle P'QR = 60°$. Considering angles at Q gives $2(90 - \alpha) + 60 = 90$, so $150 = 2\alpha$ and $\alpha = 75$.

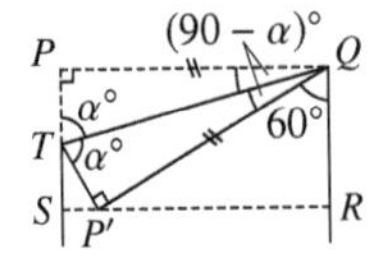

19. A As $\sin(0^2) = 0^2$ our graph must pass through the origin, so eliminating option D. As $y^2 = \sin(x^2)$, $y = \pm\sqrt{\sin(x^2)}$ and so the x-axis must be a line of symmetry, so eliminating options B and E. For some x-values, $\sin(x^2)$ will be negative and so there will be no corresponding y-value, so eliminating option C, which has y-values for every x-value. The only possible remaining option is then A and it can be checked that the graph is indeed of this form.

20. A Let $\angle ZXY = 2x°$, then the equal angles in isosceles triangle ZXY, are each $\frac{(180-2x)°}{2} = (90-x)°$. We can then find each of the angles inside triangle AZP in terms of x. Considering angles at Z gives $\angle AZP = 180° - (90-x)° = (90+x)°$. Then, as triangle OXP is isosceles, $\angle OPX = \angle OXP = 2x°$. As $\angle ZPA$ is vertically opposite $\angle OPX$, $\angle ZPA$ is also equal to $2x°$. Finally, $\angle OAY = 60°$ as triangle BAC is given to be equilateral. In triangle AZP, $60 + 90 + x + 2x = 180$, so $x = 10$ and $\angle ZXY = 2x° = 20°$.

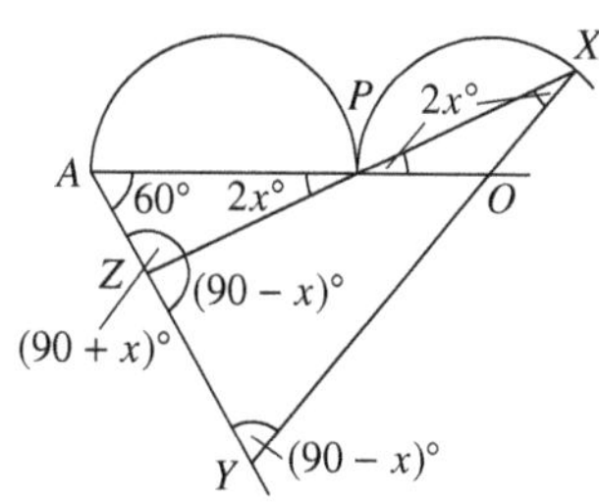

21. D Let L be the position of the ladybird on QR and let S be the position of the spider on RT each after t minutes. The shortest distance between L and S is along the straight line which is the hypotenuse of the right-angled triangle LRS. The distance QL is $30t$ cm, so the distance LR is $(1000-30t)$ cm. Also, the distance RS is $40t$ cm. So, $LS^2 = (1000-30t)^2 + (40t)^2$ which expands and simplifies to $2500(t^2 - 24t + 400) = 2500((t-12)^2 + 256)$. The distance from L to S is shortest when $t = 12$ and is $\sqrt{2500 \times 256}$ cm $= 800$ cm $= 8$ m.

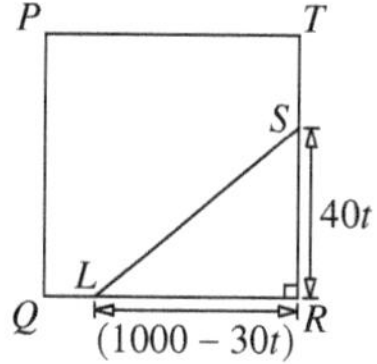

22. C At the start, taking $n = 0$ gives $(0-2019)f(0) - f(2019-0) = 2019$. Then $f(2019) = -2019(1 + f(0))$. To find $f(0)$, let $n = 2019$, then $(2019-2019)f(2019) - f(2019-2019) = 2019$, so $0 - f(0) = 2019$ and $f(0) = -2019$. Then we have $f(2019) = -2019(1 + (-2019)) = 2018 \times 2019$.

23. B When the single plane cut is made through the cube, it passes through points Y, T, V and W and also points U and X which are midpoints of two of the remaining edges of the cube as shown. The cross-section is then a regular hexagon. As the side-length of the cube is 2, the distance between the midpoints of adjacent edges is $\sqrt{2}$. This is the length of each edge of the hexagon.The hexagon can be split into six equilateral triangles and so the area of the hexagon is $6 \times \frac{1}{2} \times \sqrt{2} \times \sqrt{2} \times \sin 60° = 6 \times \frac{\sqrt{3}}{2} = 3\sqrt{3}$.

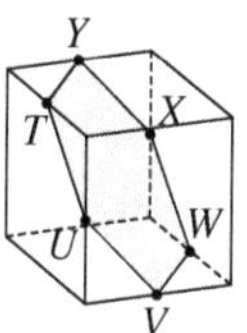

24. E Each of x, y and z has the same form which is $\sqrt{a-b} - \sqrt{a+b}$ for some a and b. Squaring this expression gives $(a-b) - 2\sqrt{a+b} \times \sqrt{a-b} + (a+b) = 2a - 2\sqrt{(a^2-b^2)}$. Applying this to x with $a = 12$ and $b = 3\sqrt{7}$ gives $x^2 = 24 - 2\sqrt{81} = 6$. Similarly we can calculate that $y^2 = 12$ and $z^2 = 2$. This gives us that $x^2y^2z^2 = 6 \times 12 \times 2 = 144$. From the initial expressions $x < 0$, $y < 0$ and $z > 0$, so $xyz > 0$ and therefore $xyz = 12$.

25. A Let the centres of the circles and the square be O_1, O_2 and X respectively. Let P be the point on the circle with centre O_1 which is a vertex of the square. Then $O_1P = 1$ and $O_1O_2 = 1$ so $O_1X = \frac{1}{2}$. Let $XP = k$, so the area of the square will be $2k^2$. As O_1X is parallel to the top edge of the square and XP goes from the centre of the square to a vertex, angle $O_1XP = 135°$. Using the cosine rule on triangle O_1XP gives $1^2 = (\frac{1}{2})^2 + k^2 - 2 \times \frac{1}{2} \times k \times \cos 135°$. As $\cos 135° = -\cos 45° = -\frac{1}{\sqrt{2}}$, this simplifies to $4k^2 + 2\sqrt{2}k - 3 = 0$. Since we know $k > 0$, $k = \frac{-1+\sqrt{7}}{2\sqrt{2}}$. So $\sqrt{2}k = \frac{-1+\sqrt{7}}{2}$ and the area of the square is $2k^2 = 2 - \frac{\sqrt{7}}{2}$.

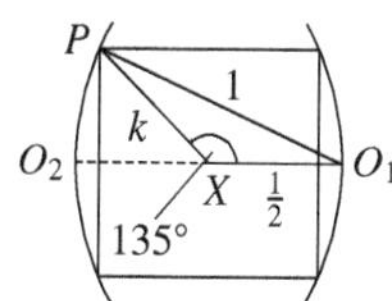

2.11.2 Participation

Participation in the Senior Mathematical Challenge

Year of competition	2015	2016	2017	2018	2019
Centres placing orders	2,223	2,253	2,270	2,290	2,144
Centres returning scripts	2,147	2,190	2,194	2,190	2,060
Papers ordered	110,500	112,350	109,500	108,100	105,670
Scripts returned	82,932	83,973	82,024	79,850	77,828

Percentage change from previous year

Year of competition	2016	2017	2018	2019
Centres placing orders	+1.3%	+0.8%	+0.9%	-6.37%
Centres returning scripts	+2.0%	+0.2%	-0.2%	-5.93%
Papers ordered	+1.7%	-2.5%	-1.3%	-2.24%
Scripts returned	+1.3%	-2.3%	-2.7%	-2.53%

2.11.3 Student performance

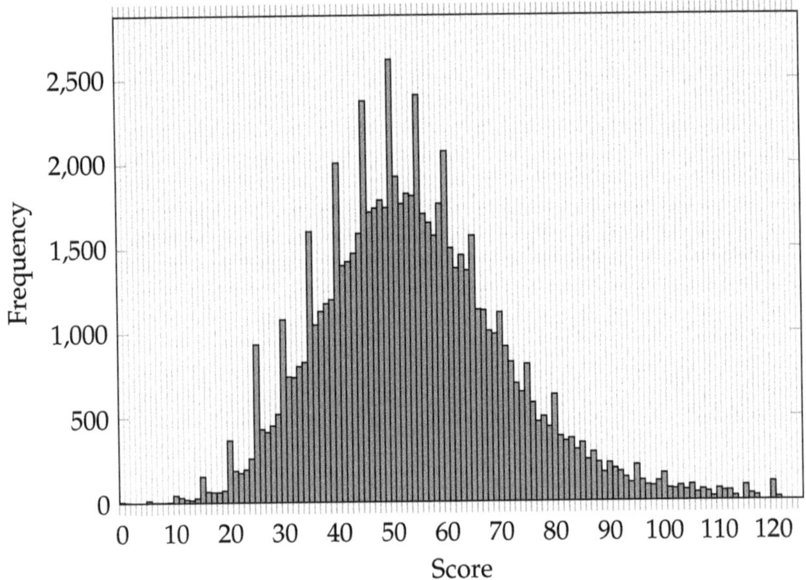

Answers given

The table below shows, for each question, the percentage of students giving each answer. The correct answer is underlined.

Question	A	B	C	D	E	Blank	Ambiguous
1	5	1	1	89	2	2	0
2	2	87	1	1	5	3	0
3	71	8	12	4	2	4	0
4	11	65	6	2	8	7	0
5	3	52	12	29	1	3	0
6	8	6	11	54	5	16	0
7	6	2	3	74	12	3	0
8	4	70	8	3	4	10	0
9	30	4	2	54	5	4	0
10	10	12	36	17	7	17	0
11	8	23	24	3	5	37	0
12	4	17	14	6	23	36	0
13	5	8	8	43	4	31	0
14	7	16	25	6	10	37	0
15	4	18	6	10	15	48	0
16	12	4	5	3	2	74	0
17	4	10	6	2	28	50	0
18	7	10	6	7	12	58	0
19	9	13	15	2	5	56	0
20	7	4	9	3	2	75	0
21	3	3	6	8	11	68	0
22	8	3	7	2	2	78	0
23	3	10	3	8	3	73	0
24	3	3	3	3	6	82	0
25	9	2	2	5	5	76	0

Mean scores, award boundaries and follow-on thresholds

To recognise the highest performers, we award the top-scoring 60 % of participants Bronze, Silver and Gold certificates in the ratio 3 : 2 : 1, and we invite around 1000 of the very highest performers to take part in the British Mathematical Olympiad Round 1 and around 6000 to take part in the Senior Kangaroo.
The table below shows the minimum score needed to obtain the corresponding award or follow-on round qualification for each of the previous five years of the competition.

Year of competition	2015	2016	2017	2018	2019
British Math. Olympiad Qual.	104	106	104	102	100
Senior Kangaroo Qual.	86	90	85	83	76
Gold	82	88	86	83	76
Silver	65	71	71	67	61
Bronze	50	55	57	53	49
Mean score	56.7	61.2	62.1	58	54.1

Comments from the Problems Group

The SMC19 average score of 54 was slightly down from the previous year (58). We were very pleased, however, that over 50% of participants answered the first nine questions correctly with some questions showing very high success rates (the first two were answered correctly by almost 90%).
As always the later questions were attempted by relatively few participants. Extended solutions to the SMC questions, with extension problems for further investigation, are available at:
https://www.ukmt.org.uk/sites/default/files/ukmt/senior-mathematical-challenge/SMC_2019_Extended_Solutions.pdf

2.12 Senior Kangaroo

2.12.1 Questions and solutions

SENIOR KANGAROO

Friday 29 November 2019

a member of the Association Kangourou sans Frontières

England & Wales: Year 13 or below
Scotland: S6 or below
Northern Ireland: Year 14 or below

INSTRUCTIONS

1. Do not open the paper until the invigilator tells you to do so.
2. Time allowed: **60 minutes**.
   No answers, or personal details, may be entered after the allowed time is over.
3. The use of blank or lined paper for rough working is allowed; **squared paper, calculators and measuring instruments are forbidden**.
4. **Use a B or an HB non-propelling pencil** to record your answer to each problem as a three-digit number from 000 to 999.
   Pay close attention to the example on the Answer Sheet that shows how to code your answers.
5. **Do not expect to finish the whole paper in the time allowed.** The questions in this paper have been arranged in approximate order of difficulty with the harder questions towards the end. You are not expected to complete all the questions during the time. You should bear this in mind when deciding which questions to tackle.
6. **Scoring rules:**
   5 marks are awarded for each correct answer;
   There is no penalty for giving an incorrect answer.
7. **The questions on this paper are designed to challenge you to think, not to guess.** You will gain more marks, and more satisfaction, by doing one question carefully than by guessing lots of answers. This paper is about solving interesting problems, not about lucky guessing.

Enquiries about the Senior Kangaroo should be sent to:

UK Mathematics Trust, School of Mathematics, University of Leeds, Leeds LS2 9JT

☎ 0113 343 2339 enquiry@ukmt.org.uk www.ukmt.org.uk

1. What is the sum of all the factors of 144?

SOLUTION **403**

The factor pairs of 144 are 1, 144; 2, 72; 3, 48; 4, 36; 6, 24; 8, 18; 9, 16 and 12 (squared). Their sum is 403.

2. When I noticed that $2^4 = 4^2$, I tried to find other pairs of numbers with this property. Trying 2 and 16, I realised that 2^{16} is larger than 16^2. How many times larger is 2^{16}?

SOLUTION **256**

$\frac{2^{16}}{16^2}=\frac{2^{16}}{(2^4)^2}=\frac{2^{16}}{2^8}=2^8=256$

3. The two diagonals of a quadrilateral are perpendicular. The lengths of the diagonals are 14 and 30. What is the area of the quadrilateral?

SOLUTION **210**

Label the quadrilateral $ABCD$ and let $AC = 14$ and $BD = 30$.
Let M be the intersection of AC and BD.
Let $AM = a$, $BM = b$, $CM = c$ and $DM = d$.
Then the sum of the areas is $\frac{1}{2}\times(ab + ad + cb + cd) = \frac{1}{2}\times(a + c)\times(b + d) = \frac{1}{2}\times14\times30 = 210$.

4. The integer n satisfies the inequality $n + (n + 1) + (n + 2) + \cdots + (n + 20) > 2019$.

What is the minimum possible value of n?

SOLUTION **087**

We solve the inequality $n + (n + 1) + (n + 2) + ... + (n + 20) > 2019$
Therefore, $21n + 210 > 2019$, i.e. $21n > 1809$ and $7n > 603$.
Therefore, $n > \frac{603}{7} = 86.1\ldots$, so n must be at least 87.

5. Identical regular pentagons are arranged in a ring. The partially completed ring is shown in the diagram. Each of the regular pentagons has a perimeter of 65. The regular polygon formed as the inner boundary of the ring has a perimeter of P. What is the value of P?

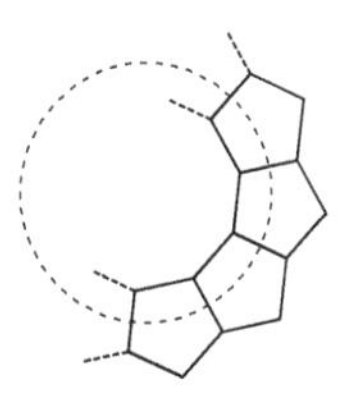

SOLUTION **130**

Let the regular N-gon at the centre of the figure have interior angles of size x degrees.
The interior angle of a pentagon is $108°$. By angles at a point we have $x + 2 \times 108 = 360$, so $x = 144$.
The exterior angle of the N-gon is $180 - 144 = 36$. Therefore, the N-gon has $\frac{360}{36} = 10$ sides.
As each side has length $\frac{65}{5} = 13$, the perimeter is $10 \times 13 = 130$.

6. For natural numbers a and b we are given that $2019 = a^2 - b^2$. It is known that $a < 1000$. What is the value of a?

SOLUTION **338**

We can write $2019 = (a + b)(a - b)$. The integers $a + b$ and $a - b$ must be a factor pair of 2019. There are two such factor pairs: $2019, 1$ and $673, 3$. These yield $(a, b) = (1010, 1009)$ and $(a, b) = (338, 335)$ respectively. As the answer must be at most 999, we conclude that $a = 338$.

7. How many positive integers n exist such that both $\frac{n+1}{3}$ and $3n + 1$ are three-digit integers?

SOLUTION **012**

For $\frac{n+1}{3}$ to be a three-figure integer we require $99 < \frac{n+1}{3} < 999$.
This simplifies to $297 < n + 1 < 2997$, that is $296 < n < 2996$.
For $3n + 1$ to be a three-figure integer we require $99 < 3n + 1 < 999$.
This simplifies to $98 < 3n < 998$, that is $\frac{98}{3} < n < \frac{998}{3}$.
These inequalities are simultaneously solved when $296 < n < \frac{998}{3} = 332\frac{2}{3}$.
For every integer value of n between 297 and 332 it is clear that $3n + 1$ will be a three-figure integer. However, $\frac{n+1}{3}$ will only be an integer for those values of $n + 1$ which are divisible by 3. These are 299, 302, 305, 308, 311, 314, 317, 320, 323, 326, 329 and 332. There are 12 numbers in this list.

8. The function $J(x)$ is defined by:

$$J(x) = \begin{cases} 4 + x & \text{for } x \leq -2, \\ -x & \text{for } -2 < x \leq 0, \\ x & \text{for } x > 0. \end{cases}$$

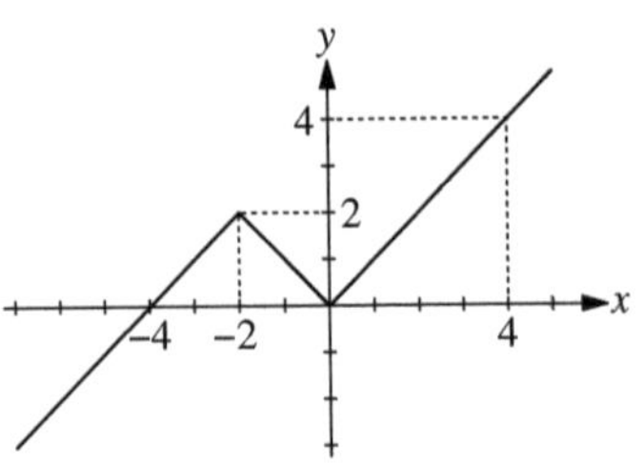

How many distinct real solutions has the equation $J(J(J(x))) = 0$?

Solution **004**

The only solutions to $J(x) = 0$ are $x = 0, -4$.
Since $J(0) = 0$, both will also be solutions of $J(J(J(x))) = 0$.
Any solution to $J(x) = -4$ will also be a solution to $J(J(x)) = 0$. The only solution to $J(x) = -4$ is $x = -8$. Since $J(x) = 0$, $x = -8$ is also a solution of $J(J(J(x))) = 0$.
Any solution to $J(x) = -8$ will also be a solution to $J(J(J(x))) = 0$. The only solution to $J(x) = -8$ is $x = -12$.
Therefore, there are four distinct solutions, $x = 0, -4, -8$ and -12.

9. What is the smallest three-digit number K which can be written as $K = a^b + b^a$, where both a and b are one-digit positive integers?

Solution **100**

As the problem is symmetrical in a, b we assume $a \leq b$ without loss of generality.
If $a = 1$ then the maximum value of $a^b + b^a$ is $1^9 + 9^1 = 1 + 9 = 10$. This is not a three-digit number, so cannot be a value of K.
If $a = 2$ then possible values for K include $2^9 + 9^2 = 512 + 81 = 593$, $2^8 + 8^2 = 256 + 64 = 320$, $2^7 + 7^2 = 128 + 49 = 177$ and $2^6 + 6^2 = 64 + 36 = 100$.
As 100 can be attained then 100 is the smallest three-digit number K.

10. What is the value of $\sqrt{13+\sqrt{28+\sqrt{281}}} \times \sqrt{13-\sqrt{28+\sqrt{281}}} \times \sqrt{141+\sqrt{281}}$?

SOLUTION **140**

$$\sqrt{13+\sqrt{28+\sqrt{281}}} \times \sqrt{13-\sqrt{28+\sqrt{281}}} \times \sqrt{141+\sqrt{281}}$$

$$= \sqrt{13^2 - \left(\sqrt{28+\sqrt{281}}\right)^2} \times \sqrt{141+\sqrt{281}}$$

$$= \sqrt{169 - \left(28+\sqrt{281}\right)} \times \sqrt{141+\sqrt{281}}$$

$$= \sqrt{141-\sqrt{281}} \times \sqrt{141+\sqrt{281}} = \sqrt{\left(141-\sqrt{281}\right) \times \left(141+\sqrt{281}\right)}$$

$$= \sqrt{141^2 - 281} = \sqrt{141^2 - 282 + 1} = \sqrt{(141-1)^2} = \sqrt{140^2} = 140$$

11. In the triangle ABC the points M and N lie on the side AB such that $AN = AC$ and $BM = BC$.

We know that $\angle MCN = 43°$.

Find the size in degrees of $\angle ACB$.

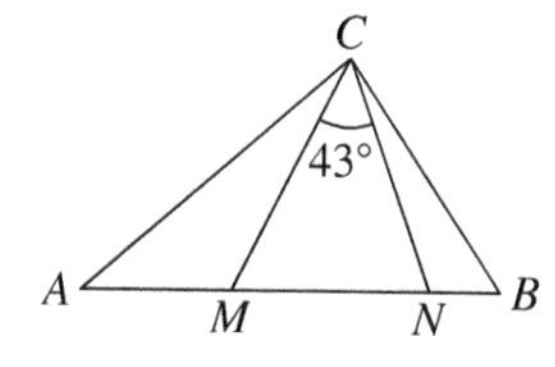

SOLUTION **094**

Let $\angle ACM = x°$ and $\angle BCN = y°$.
Using the base angles property of isosceles triangles ACN and BCM, we have $\angle ANC = 43 + x$ and $\angle BMC = 43 + y$.
In triangle CMN, $43 + (43 + x) + (43 + y) = 180$.
Therefore, $\angle ACB = x + 43 + y = 94$.

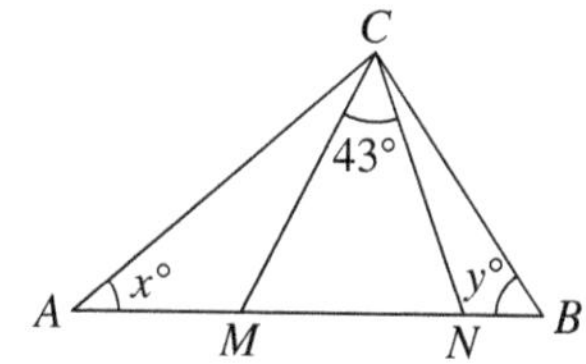

12. What is the value of $A^2 + B^3 + C^5$, given that:

$$A = \sqrt[3]{16\sqrt{2}}$$
$$B = \sqrt{9\sqrt[3]{9}}$$
$$C = [(\sqrt[5]{2})^2]^2$$

SOLUTION **105**

$$A^2 = \left(\sqrt[3]{16 \times \sqrt{2}}\right)^2 = \left(2^4 \times 2^{\frac{1}{2}}\right)^{\frac{2}{3}} = \left(2^{\frac{9}{2}}\right)^{\frac{2}{3}} = 2^{\frac{18}{6}} = 2^3 = 8$$

$$B^3 = \sqrt{9 \times \sqrt[3]{9}}^{\,3} = \left(9 \times 9^{\frac{1}{3}}\right)^{\frac{3}{2}} = \left(9^{\frac{4}{3}}\right)^{\frac{3}{2}} = 9^{\frac{12}{6}} = 9^2 = 81$$

$$C^5 = \left(\left(\sqrt[5]{2}^2\right)^2\right)^5 = \left(2^{\frac{1}{5}}\right)^{2\times2\times5} = 2^{\frac{20}{5}} = 2^4 = 16$$

$$A^2 + B^3 + C^5 = 8 + 81 + 16 = 105$$

13. The real numbers a and b, where $a > b$, are solutions to the equation $3^{2x} - 10\times3^{x+1} + 81 = 0$. What is the value of $20a^2 + 18b^2$?

SOLUTION **198**

In the equation $3^{2x} - 10 \times 3^{x+1} + 81 = 0$, replace 3^x with y. The equation becomes $y^2 - 10 \times 3 \times y + 81 = 0$. This factorises as $(y - 3)(y - 27) = 0$ with solutions $y = 3, 27$. This means $3^x = 3$ or $3^x = 27$. The x-values are $1, 3$ respectively, so $a = 3$ and $b = 1$. The value of $20a^2 + 18b^2 = 20 \times 9 + 18 \times 1 = 198$.

14. A number N is the product of three distinct primes. How many distinct factors does N^5 have?

SOLUTION **216**

Let the three distinct prime factors of N be p, q and r. Therefore, $N^5 = p^5 \times q^5 \times r^5$. Each factor of N^5 may be written as $p^a \times q^b \times r^c$, where $a, b, c \in \{0, 1, 2, 3, 4, 5\}$. Since there are 6 choices for the value of each of a, b, c there are $6 \times 6 \times 6 = 216$ distinct factors of N^5.

15. Five Bunchkins sit in a horizontal field. No three of the Bunchkins are sitting in a straight line. Each Bunchkin knows the four distances between her and each of the others. Each Bunchkin calculates and then announces the total of these distances. These totals are 17, 43, 56, 66 and 76. A straight line is painted joining each pair of Bunchkins. What is the total length of paint required?

SOLUTION **129**

Each line's length will be announced twice; once by each of the two Bunchkins at its ends. By adding up the total of the numbers announced we will therefore include the length of each line exactly twice.
The total length of paint required is $\frac{1}{2} \times (17 + 43 + 56 + 66 + 76) = \frac{258}{2} = 129$.

16. The real numbers x and y satisfy the equations:

$$xy - x = 180 \quad \text{and} \quad y + xy = 208.$$

Let the two solutions be (x_1, y_1) and (x_2, y_2).
What is the value of $x_1 + 10y_1 + x_2 + 10y_2$?

SOLUTION **317**

Subtracting the two equations yields $y + x = 28$. Substituting $y = 28 - x$ into $xy - x = 180$ leads to the quadratic equation $0 = x^2 - 27x + 180$. This has solutions 12, 15. The solution sets are therefore (15, 13) and (12, 16).
The value of $x_1 + 10y_1 + x_2 + 10y_2$ is $15 + 10 \times 13 + 12 + 10 \times 16 = 317$.

17. In triangle ABC, $\angle BAC$ is 120°. The length of AB is 123. The point M is the midpoint of side BC. The line segments AB and AM are perpendicular.
What is the length of side AC?

SOLUTION **246**

Extend the line BA. Draw a line through C, parallel to MA, meeting the extended line BA at point N. By the intercept theorem, $BA = AN = 123$, because $BM = MC$. In triangle NAC, $\cos 60 = \frac{1}{2} = \frac{123}{AC}$. Therefore, $AC = 246$.

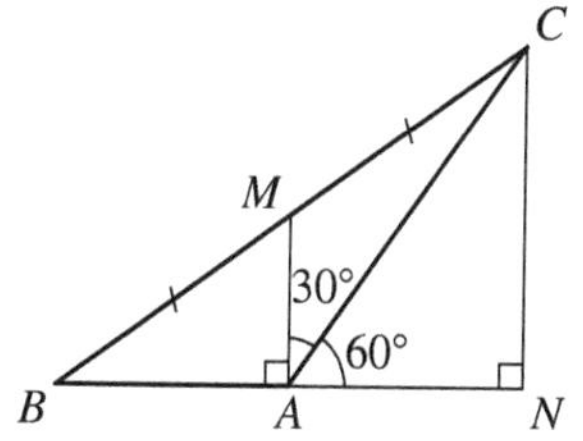

18. An integer is said to be *chunky* if it consists only of non-zero digits by which it is divisible when written in base 10.
For example, the number 936 is Chunky since it is divisible by 9, 3 and 6.
How many chunky integers are there between 13 and 113?

SOLUTION **014**

For a two-digit number N = "ab" we may write $N = 10a + b$.
If N is Chunky, then N will be divisible by a and therefore $b = N - 10a$ will also be divisible by a.
It is therefore sufficient to check only those two-digit numbers which have a units digit divisible by their tens digit. Following checking, 15, 22, 24, 33, 36, 44, 48, 55, 66, 77, 88 and 99 are the only two-digit Chunky numbers (excluding 11 and 12, which are not under consideration). Of those three-digit numbers under consideration, only 111 and 112 are Chunky. The answer is 14.

19. The square $ABCD$ has sides of length 105. The point M is the midpoint of side BC. The point N is the midpoint of BM. The lines BD and AM meet at the point P. The lines BD and AN meet at the point Q.
What is the area of triangle APQ?

SOLUTION **735**

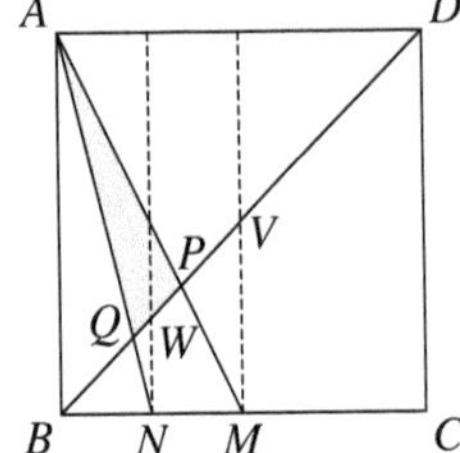

Let V be the centre of the square $ABCD$. Let W be the intersection between BD and the line through N parallel to AB.
Triangles APB and MPV are similar, with $BP : PV = AB : MV = \frac{1}{2}$. Therefore, $BP = \frac{2}{3} \times BV = \frac{2}{3} \times \frac{1}{2} \times 105\sqrt{2} = 35\sqrt{2}$.
Similarly, triangles AQB and NQW are similar, with $BQ : QW = AB : NW = \frac{1}{4}$. Therefore, $BQ = \frac{4}{5} \times BW = \frac{4}{5} \times \frac{1}{4} \times 105\sqrt{2} = 21\sqrt{2}$.
The area of APQ is $\frac{1}{2} \times QP \times VA = \frac{1}{2} \times \left(35\sqrt{2} - 21\sqrt{2}\right) \times \frac{1}{2} \times 105\sqrt{2} = \frac{1}{2} \times 14\sqrt{2} \times \frac{1}{2} \times 105\sqrt{2} = 735$.

20. Each square in this cross-number can be filled with a non-zero digit such that all of the conditions in the clues are fulfilled. The digits used are not necessarily distinct. What is the answer to 3 ACROSS?

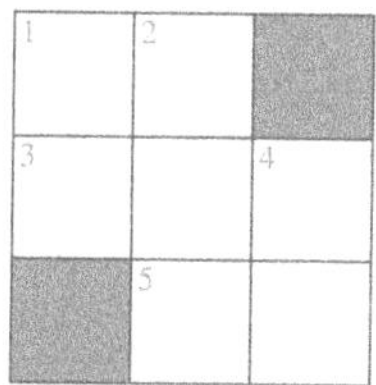

ACROSS	DOWN
1. A composite factor of 1001	1. One more than a prime, one less than a prime
3. Not a palindrome	2. A multiple of 9
5. pq^3 where p, q prime and $p \neq q$	4. p^3q using the same p, q as 5 ACROSS

SOLUTION **295**

1 Across may be either 77 or 91. The only possibility for 1 Down with 7 or 9 as its first digit is 72. So 1 Across is 77 and 1 Down is 72.

In the clues for 5 Across and 4 Down we see that p, q must be 2, 3 in some order, since if any larger prime were used then pq^3 and qp^3 would not both be two-digit. Therefore, 5 Across and 4 Down are $3 \times 2^3 = 24$ and $2 \times 3^3 = 54$ in some order. We know that 3 Across is not a palindrome (so may not end in a 2). Therefore, 5 Across is 24 and 4 Down is 54.

The only three-digit multiples of 9 beginning with a 7 are 702 and 792. As every digit in the completed crossnumber must be non-zero we have 2 Down is 792 and 3 Across is 295.

2.12.2 Participation

Participation in the Senior Kangaroo

Year of competition	2015	2016	2017	2018	2019
Centres with invitees	1,071	1,234	1,261	1,203	1,277
Centres returning scripts	972	1,098	1,143	1,093	1,164
Invitees	3,961	5,445	6,230	5,508	6,312
Scripts returned	2,960	4,079	4,623	4,066	4,839

Percentage change from previous year

Year of competition	2016	2017	2018	2019
Centres with invitees	+15.2%	+2.2%	-4.6%	+6.15%
Centres returning scripts	+13.0%	+4.1%	-4.4%	+6.45%
Invitees	+37.5%	+14.4%	-11.6%	+14.6%
Scripts returned	+37.8%	+13.3%	-12.0%	+19.01%

2.12.3 Student performance

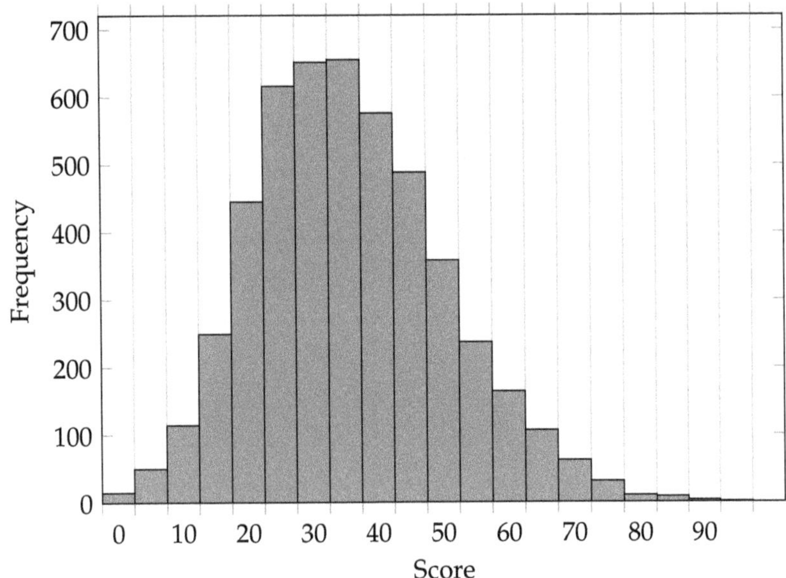

Answers given

The table below shows, for each question, the percentage of students giving the correct answer and the percentage leaving the question blank.

Question	Correct	Blank
1	35	1
2	91	1
3	62	13
4	62	2
5	78	6
6	36	40
7	25	12
8	18	32
9	71	10
10	32	45
11	39	37
12	47	27
13	39	46
14	12	38
15	15	63
16	26	55
17	6	71
18	15	54
19	3	79
20	3	70

Mean scores and award boundaries

The table shows, for each of the past five years of the competition, the minimum score needed to obtain a Certificate of Merit and the mean score.

Year of competition	2015	2016	2017	2018	2019
Certificate of Merit	50	55	40	35	45
Mean score	39.5	42.3	28.5	24.2	35.8

2.13 British Mathematical Olympiad Round 1

2.13.1 Questions and solutions

United Kingdom
Mathematics Trust

BRITISH MATHEMATICAL OLYMPIAD
ROUND 1
Friday 29 November 2019

INSTRUCTIONS

1. Time allowed: $3\frac{1}{2}$ hours.
2. Full written solutions – not just answers – are required, with complete proofs of any assertions you may make. Marks awarded will depend on the clarity of your mathematical presentation. Work in rough first, and then write up your best attempt. Do not hand in rough work.
3. One complete solution will gain more credit than several unfinished attempts. It is more important to complete a small number of questions than to try all the problems.
4. Each question carries 10 marks. However, earlier questions tend to be easier. In general you are advised to concentrate on these problems first.
5. The use of rulers, set squares and compasses is allowed, but calculators and protractors are forbidden.
6. Start each question on a fresh sheet of paper. Write on one side of the paper only. On each sheet of working write the number of the question in the top left hand corner and your **Participant ID**, and **UKMT Centre Number** in the top right hand corner.
7. Complete the cover sheet provided and attach it to the front of your script, followed by your solutions in question number order.
8. Staple all the pages neatly together in the top left hand corner.
9. To accommodate candidates sitting in other time zones, please do not discuss the paper on the internet until 8am GMT on Saturday 30 November when the solutions video will be released at https://bmos.ukmt.org.uk
10. **Do not turn over until told to do so.**

Enquiries about the British Mathematical Olympiad should be sent to:

UK Mathematics Trust, School of Mathematics, University of Leeds, Leeds LS2 9JT
☎ 0113 343 2339 enquiry@ukmt.org.uk www.ukmt.org.uk

5

BMO Round 1

Problem 1 (Proposed by Nick MacKinnon) *Show that there are at least three prime numbers p less than 200 for which $p+2$, $p+6$, $p+8$ and $p+12$ are all prime. Show also that there is only one prime number q for which $q+2$, $q+6$, $q+8$, $q+12$ and $q+14$ are all prime.*

Solution by Yuka Machino: When p is any of 5, 11 or 101, all of p, $p+2$, $p+6$, $p+8$ and $p+12$ are prime.

If we look at q, $q+2$, $q+6$, $q+8$, $q+12$ and $q+14$ modulo five, then these become

$$q, q+2, q+1, q+3, q+2, \text{ and } q+4.$$

Now, modulo five, the five integers $q, q+1, q+2, q+3, q+4$ are all distinct so one of them is divisible by five. As they all have to be prime, one of them must equal five. But $q > 0$ so $q+6 > 5$ and so there are only two possibilities: $q = 5$ or $q+2 = 5$.

If $q+2 = 5$, then $q+6 = 9$ which is not prime. However, if $q = 5$, then all of q, $q+2$, $q+6$, $q+8$, $q+12$ and $q+14$ are prime and so $q = 5$ is the unique prime number satisfying the property in the question.

Problem 2 (Proposed by Tom Bowler) *A sequence of integers $a_1, a_2, a_3, \ldots$ satisfies the relation:*

$$4a_{n+1}^2 - 4a_n a_{n+1} + a_n^2 - 1 = 0$$

for all positive integers n. What are the possible values of a_1?

Solution by Scott Hall: We may factorise the given condition as

$$(2a_{n+1} - a_n)^2 = 1,$$

which is equivalent to

$$2a_{n+1} = a_n + 1 \quad \text{or} \quad 2a_{n+1} = a_n - 1.$$

In particular, at least one of $a_n + 1$ or $a_n - 1$ must be even, so a_n must be odd for all n. Thus, a_1 must be an odd integer.

We claim that if a_n is odd, then we may choose a_{n+1} to both satisfy the relation in the question and be odd. Indeed, if we let $a_n = 2k + 1$ (for some integer k), then

$$a_{n+1} = \frac{a_n + 1}{2} = k + 1 \quad \text{or} \quad a_{n+1} = \frac{a_n - 1}{2} = k.$$

Now, k and $k+1$ are two consecutive integers, so one of them is odd. Choosing a_{n+1} to be the one that is odd means a_{n+1} is both odd and satisfies the relation. Thus, if a_1 is an odd integer, then we may form a sequence $a_1, a_2, a_3, \ldots$ of odd integers all satisfying the relation.

Hence, the possible values of a_1 are exactly the odd integers.

Problem 3 (Proposed by Daniel Griller) *Two circles S_1 and S_2 are tangent at P. A common tangent, not through P, touches S_1 at A and S_2 at B. Points C and D, on S_1 and S_2 respectively, are outside the triangle APB and are such that P is on the line CD.*

Prove that AC is perpendicular to BD.

Solution 1 by Velian Velikov: Let O_1 and O_2 be the centres of S_1 and S_2 respectively and let lines AC and BD meet at Q. To prove that AC is perpendicular to BD, we need to show that $C\hat{Q}D = 90°$.

Circles S_1 and S_2 touch at P, so P lies on line O_1O_2. Since AB is tangent to both S_1 and S_2, both $O_1\hat{A}B$ and $A\hat{B}O_2$ are right angles.

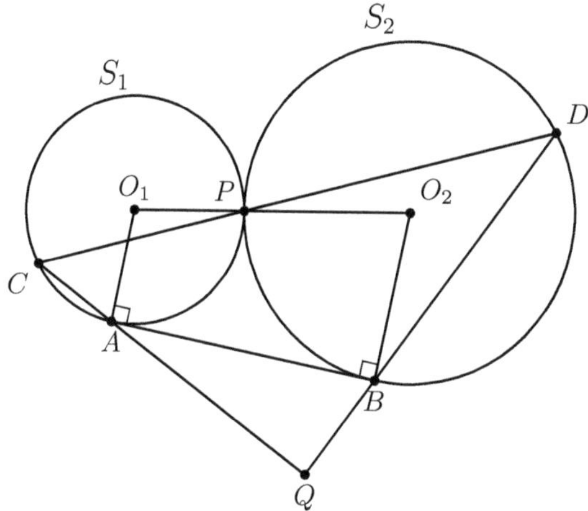

The angles in quadrilateral ABO_2O_1 sum to $360°$, so $P\hat{O}_1A + B\hat{O}_2P = 180°$. But, as the angle at the centre of a circle is twice the angle at the circumference, $P\hat{O}_1A = 2P\hat{C}A$ and $B\hat{O}_2P = 2B\hat{D}P$. Hence $P\hat{C}A + B\hat{D}P = 90°$.

Finally, consider triangle CQD: $D\hat{C}Q + Q\hat{D}C = P\hat{C}A + B\hat{D}P = 90°$. Hence $C\hat{Q}D = 90°$, so AC is perpendicular to BD.

Solution 2 by Shao Yuan (Victor) Lin: Let the common tangent to S_1 and S_2 which passes through P meet line AB at the point M and let lines AC and BD meet at Q. Segments MA and MP are both tangents from M to S_1 so $MA = MP$. Similarly $MP = MB$. Thus, M is the centre of the circle with diameter AB and so $B\hat{P}A$ is a right angle.

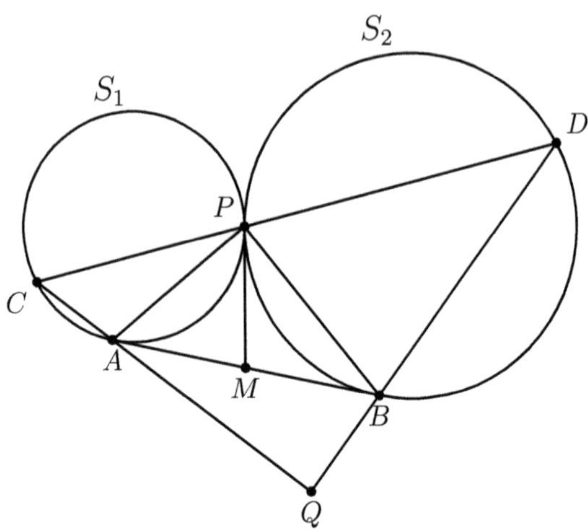

Considering right-angled triangle ABP, we have $P\hat{A}B + A\hat{B}P = 90°$. Using the alternate segment theorem, $P\hat{C}A = P\hat{A}B$ and $B\hat{D}P = A\hat{B}P$, and hence $P\hat{C}A + B\hat{D}P = 90°$.

Finally, considering triangle CQD: $D\hat{C}Q + Q\hat{D}C = P\hat{C}A + B\hat{D}P = 90°$. Hence $C\hat{Q}D = 90°$, and so AC is indeed perpendicular to BD.

Problem 4 (Proposed by Daniel Griller) *There are* 2019 *penguins waddling towards their favourite restaurant. As the penguins arrive, they are handed tickets numbered in ascending order from* 1 *to* 2019*, and told to join the queue. The first penguin starts the queue. For each* $n > 1$ *the penguin holding ticket number* n *finds the greatest* $m < n$ *which divides* n *and enters the queue directly behind the penguin holding ticket number* m*. This continues until all* 2019 *penguins are in the queue.*

(a) *How many penguins are in front of the penguin with ticket number* 2*?*

(b) *What numbers are on the tickets held by the penguins just in front of and just behind the penguin holding tickets* 33*?*

Solution by Sasha Walker: We will write 'penguin n' for the penguin with ticket number n. We will write each number as the product of its prime factors in descending order, so, for example, 84 would be written as $7 \cdot 3 \cdot 2 \cdot 2$.

If n is written as $a_1 \cdot a_2 \cdot \ldots \cdot a_k$, then the greatest $m < n$ which divides n is $a_1 \cdot a_2 \cdot \ldots \cdot a_{k-1}$ (if n is prime, so written as n, then m is 1). Hence, penguin $n = a_1 \cdot a_2 \cdot \ldots \cdot a_k$ will enter the queue directly behind penguin $m = a_1 \cdot a_2 \cdot \ldots \cdot a_{k-1}$ (although they may end up being separated). Conversely, the penguins that will enter the queue directly behind penguin $m = a_1 \cdot a_2 \cdot \ldots \cdot a_{k-1}$ are exactly those written as $a_1 \cdot a_2 \cdot \ldots \cdot a_k$ where $a_k \leqslant a_{k-1}$. If $a_k < a'_k \leqslant a_{k-1}$, then both penguin $n = a_1 \cdot a_2 \cdot \ldots \cdot a_{k-1} \cdot a_k$ and penguin $n' = a_1 \cdot a_2 \cdot \ldots \cdot a_{k-1} \cdot a'_k$ will enter directly behind penguin m but penguin n' will enter later, so will be further forward in the queue.

Therefore, we can find the order of the queue by viewing it as a dictionary where each number is a word and the letters are its prime factors in descending order. Higher prime factors come first and if one word is the initial part of another, then the shorter word comes first.

(a) The only penguin to enter directly behind penguin 2 is penguin $2 \cdot 2 = 4$. The only penguin to enter directly behind penguin 4 is penguin $2 \cdot 2 \cdot 2 = 8, \ldots$. Thus, at the end, the penguins behind penguin 2 are exactly penguin 4, penguin $8, \ldots,$ penguin 1024 – that is, there are nine penguins behind penguin 2. Hence, the number of penguins in front of penguin 2 is $2019 - 1 - 9 = 2009$.

(b) We write 33 as $11 \cdot 3$. The only penguins to enter directly behind penguin 33 are penguins $11 \cdot 3 \cdot 3 = 99$ and $11 \cdot 3 \cdot 2 = 66$. Penguin 99 enters the queue later, so is the penguin directly behind penguin 33 in the final queue.

Let N be the number of the penguin directly in front of penguin 33 in the final queue. Penguin 33 enters directly behind penguin 11. The penguins which enters directly behind penguin 11 are $22, 33, 55, 77$ and, in the final queue their relative order is $11, 77, 55, 33, 22$. Hence, in the final queue, penguin N is between penguins 55 and 33. The penguins to enter directly behind penguin $55 = 11 \cdot 5$ are $55 \cdot 5 = 275$, $55 \cdot 3 = 165$, $55 \cdot 2 = 110$ and in the final queue, penguin 110 will be the last of those. Hence, in the final queue, penguin N is between penguins 110 and 33.

The only penguin to enter directly behind penguin $110 = 11 \cdot 5 \cdot 2$ is penguin $11 \cdot 5 \cdot 2 \cdot 2$. The only penguin to enter behind this is $11 \cdot 5 \cdot 2 \cdot 2 \cdot 2$. Now, $55 \times 2^5 = 1760 < 2019$, while $55 \times 2^6 > 2019$ so, repeating the argument just given, we obtain $N = 1760$.

Problem 5 (Proposed by Dominic Rowland) *Six children are evenly spaced around a circular table. Initially, one has a pile of $n > 0$ sweets in front of them, and the others have nothing. If a child has at least four sweets in front of them, they may perform the following move: eat one sweet and give one sweet to each of their immediate neighbours and to the child directly opposite them. An arrangement is called* perfect *if there is a sequence of moves which results in each child having the same number of sweets in front of them. For which values of n is the initial arrangement perfect?*

Solution by Reuben Mason: We will colour the six spaces black and white and label them as follows. At the start, space A has n sweets and the rest have no sweets.

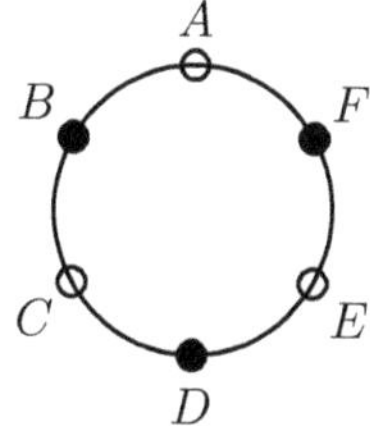

We use 'A makes a move' to mean A eats a sweet and passes a sweet to B, D and F (and similarly define 'B makes a move', ...). Consider the effect of A making a move: A loses four sweets, the black spaces each gain a sweet and the other two are unchanged. Consider this move modulo four: the number of sweets in each white space is unchanged modulo four, while the number of sweets in each of the black spaces increases by one modulo four. Hence, the difference between the number of sweets of two spaces of the same colour is invariant modulo four. At the start, the difference between A and C's number of sweets is n. If the initial arrangement is perfect, then at the end this difference will be zero and so $n \equiv 0 \pmod 4$. That is, if the initial arrangement is perfect, then n is a multiple of four.

Consider a move made by a white space: this reduces the total number of sweets in white spaces by four and increases the total number of sweets in black spaces by three. Similarly a move by a black space reduces the total number of sweets in black spaces by four and increases the total number of sweets in white spaces by three. Let there be b moves by black spaces and w moves by white spaces. Initially, there are n sweets in white spaces and no sweets in black spaces and so at the end there will be $n - 4w + 3b$ sweets in white spaces and $3w - 4b$ sweets in black spaces. If the arrangement is perfect, then these two numbers will be equal: $n - 4w + 3b = 3w - 4b$. Rearranging gives $n = 7(b + w)$ and so n is a multiple of seven.

Hence if the initial arrangement is perfect, then n is divisible by 28.

We finally show that, if n is divisible by 28, then the initial arrangement is perfect. Let $n = 28k$ where k is a positive integer: initially, the sweets are distributed as in the left diagram. If A makes $7k$ moves, then the sweets are distributed as in the middle diagram. Finally if each of B, D and E makes k moves, then the sweets will be distributed as in the right diagram. Hence the initial arrangement is perfect exactly if n is a multiple of 28.

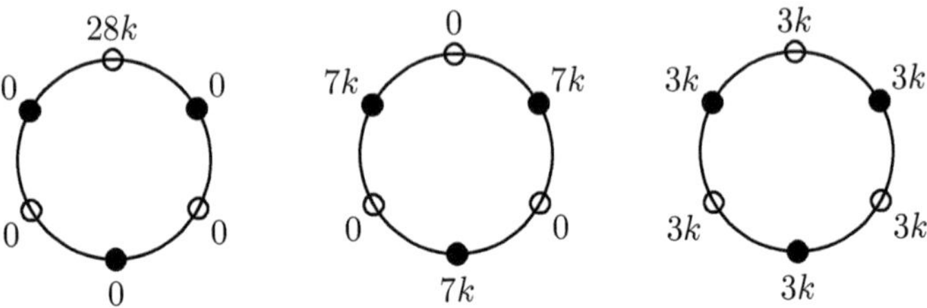

Problem 6 (Proposed by Sam Bealing) *A function f is called* good *if it assigns an integer value $f(m,n)$ to every ordered pair of integers (m,n) in such a way that for every pair of integers (m,n) we have:*

$$2f(m,n) = f(m-n, n-m) + m + n = f(m+1,n) + f(m,n+1) - 1.$$

Find all good functions.

Solution 1 by Seung Jae (Sam) Yang: We first concentrate on the left-hand equality. Taking $m = n = 0$ gives $2f(0,0) = f(0,0)$ so $f(0,0) = 0$, while taking $m = n = k$ gives $2f(k,k) = f(0,0) + 2k = 2k$, so, for all k,

$$f(k,k) = k.$$

We will now investigate how the values of $f(m,n)$, $f(m+1,n)$, $f(m,n+1)$ and $f(m+1,n+1)$ relate to each other. Firstly, rearranging $2f(m,n) = f(m+1,n) + f(m,n+1) - 1$ gives

$$[f(m+1,n) - f(m,n)] + [f(m,n+1) - f(m,n)] = 1. \tag{1}$$

Replacing m by $m+1$ and n by $n+1$ in the left-hand equality gives

$$2f(m+1,n+1) = f(m-n,n-m) + m + n + 2 = 2f(m,n) + 2,$$

so

$$f(m+1,n+1) = f(m,n) + 1. \tag{2}$$

Let $a = f(m+1,n) - f(m,n)$ and $b = f(m,n+1) - f(m,n)$. Equation (1) says that $a + b = 1$. We visualise this and equation (2) in the following diagram. The number on an arrow pointing from the point (m,n) to the point (m',n') is the value of $f(m',n') - f(m,n)$ ("the change in f when you move from (m,n) to (m',n')")

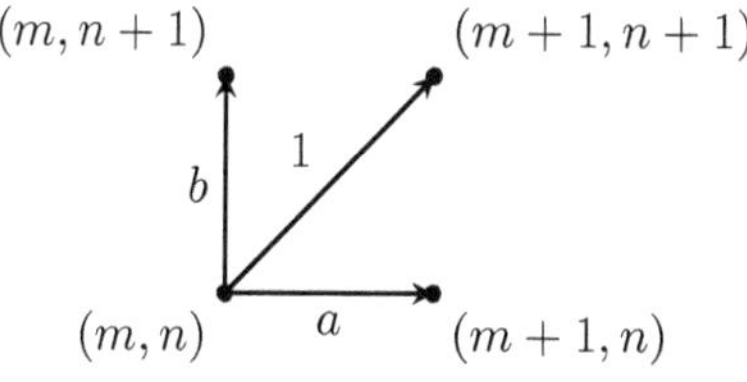

Considering the triangle of points (m,n), $(m+1,n)$ and $(m+1,n+1)$, we see that $f(m+1,n+1) - f(m+1,n) = 1 - a = b$. Similarly $f(m+1,n+1) - f(m,n+1) = 1 - b = a$. We add these to the diagram as well as the point $(m-1,n)$.

2.13.2 Participation

Participation in the British Mathematical Olympiad Round 1

Year of competition	2015–16	2016–17	2017–18	2018–19	2019–20
Centres with invitees	437	495	462	411	397
Centres returning scripts	519	566	565	516	490
Invitees	924	1,125	1,096	933	903
Scripts returned	1,549	1,751	1,828	1,583	1,560

Percentage change from previous year

Year of competition	2016–17	2017–18	2018–19	2019–20
Centres with invitees	+13.3%	-6.7%	-11.0%	-3.4%
Centres returning scripts	+9.1%	-0.2%	-8.7%	-5.03%
Invitees	+21.8%	-2.6%	-14.9%	-3.12%
Scripts returned	+13.0%	+4.4%	-13.4%	-1.45%

2.13.3 Student performance

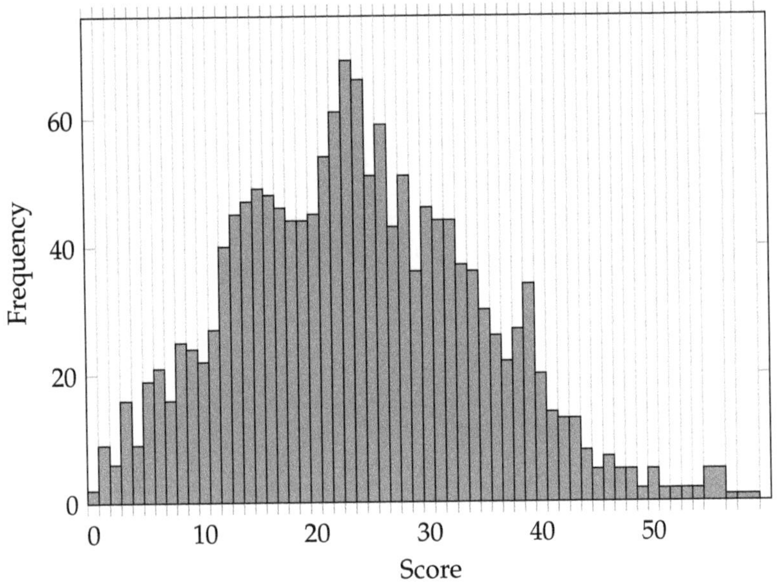

Mean scores and award boundaries

We invite around 100 top scorers to take part in the British Mathematical Olympiad Round 2. The top 25% of scorers receive a Certificate of Distinction. Of the rest, those who qualified automatically to sit the paper receive a Certificate of Qualification. We award Gold, Silver and Bronze medals to the top 100 or so candidates in the ratio 1 : 2 : 3. All medallists receive a book prize, which varies from year to year.
The table below shows the minimum score needed to obtain the corresponding award or follow-on round qualification, and the mean score.

Year of competition	2015–16	2016–17	2017–18	2018–19	2019–20
BMO2 (Y13/S6/NI14)	35	42	47	35	40
BMO2 (Y12/S5/NI13)	31	38	42	32	40
BMO2 (Y11/S4/NI12)	28	35	37	29	40
BMO2 (<Y11/S4/NI12)	23	30	35	26	40
Certificate of Distinction	23	30	28	22	32
Gold	50	52	59	48	54
Silver	40	46	50	41	45
Bronze	33	41	42	35	41
Mean score	16.3	22.9	21.2	15.8	23.9

2.14 British Mathematical Olympiad Round 2

2.14.1 Questions and solutions

United Kingdom
Mathematics Trust

BRITISH MATHEMATICAL OLYMPIAD
ROUND 2

Thursday 30 January 2020

supported by 

INSTRUCTIONS

1. Time allowed: $3\frac{1}{2}$ hours. Each question is worth 10 marks.
2. Full written solutions – not just answers – are required, with complete proofs of any assertions you may make. Marks awarded will depend on the clarity of your mathematical presentation. Work in rough first, and then draft your final version carefully before writing up your best attempt.
3. Write on one side of the paper only and start each question on a fresh sheet.
4. Rough work *should* be handed in, but should be clearly marked.
5. One or two *complete* solutions will gain far more credit than partial attempts at all four problems.
6. The use of rulers and compasses is allowed, but calculators and protractors are forbidden.
7. Staple all the pages neatly together in the top *left* hand corner, with questions 1, 2, 3, 4 in order, and the cover sheet at the front.
8. To accommodate candidates sitting in other time zones, please do not discuss any aspect of the paper on the internet until 8am GMT on Friday 31 January. Candidates sitting the paper in time zones more than 3 hours ahead of GMT must sit the paper on Friday 31 January (as defined locally).
9. In early March, twenty-four students eligible to represent the UK at the International Mathematical Olympiad will be invited to attend the training session to be held at Trinity College, Cambridge (31 March–5 April 2020). At the training session, students sit a pair of IMO-style papers and some students will be selected for further training and selection examinations. The UK Team of six for this year's IMO (to be held in St Petersburg, Russia 8–18 July 2020) will then be chosen.
10. **Do not turn over until told to do so.**

Enquiries about the British Mathematical Olympiad should be sent to:

UK Mathematics Trust, School of Mathematics, University of Leeds, Leeds LS2 9JT

☎ 0113 343 2339 enquiry@ukmt.org.uk www.ukmt.org.uk

8

BMO Round 2

Problem 1 (Proposed by Sam Maltby) *A sequence $a_1, a_2, a_3, \ldots$ has $a_1 > 2$ and satisfies:*

$$a_{n+1} = \frac{a_n(a_n - 1)}{2}$$

for all positive integers n. For which values of a_1 are all the terms of the sequence odd integers?

Solution by Henry Jaspars: First notice that if $a_1 = 3$, then the sequence is the constant sequence of 3's and so consists solely of odd integers.

From now on we assume that $a_1 > 3$. Let $b_n = a_n - 3$. The sequence $b_1, b_2, b_3, \ldots$ has $b_1 > 0$ and satisfies

$$\begin{aligned} b_{n+1} = a_{n+1} - 3 &= \frac{a_n(a_n - 1)}{2} - 3 \\ &= \frac{(b_n + 3)(b_n + 2)}{2} - 3 = \frac{b_n(b_n + 5)}{2}. \end{aligned}$$

If all of $a_1, a_2, a_3, \ldots$ are odd, then all of $b_1, b_2, b_3, \ldots$ are even. Suppose that the highest power of 2 dividing b_1 is 2^k. Then $b_1 + 5$ is odd and so the highest power of 2 dividing b_2 is 2^{k-1}. Repeating this we obtain that b_k is odd, which is a contradiction.

Hence, the only value of a_1 for which all of $a_1, a_2, a_3, \ldots$ are odd is $a_1 = 3$.

Problem 2 (Proposed by Sam Bealing) *Describe all collections S of at least four points in the plane such that no three points are collinear and such that every triangle formed by three points in S has the same circumradius.*

(The circumradius of a triangle is the radius of the circle passing through all three of its vertices)

Solution by Sarah Henderson: We will show that there are exactly two types of collection that work: a set of at least four points all lying on a circle or a set of four points consisting of a non-right angled triangle together with its orthocentre. *In any triangle, the three altitudes all concur at a point which is called the* orthocentre.

Certainly any collection of four or more points all lying on a circle works: each triangle from this collection has the same circumcircle.

Now suppose that not all the points in S lie on the same circle: let ABC be a triangle in S and P a point of S not on circle ABC. P lies on a circle through A and B with the same radius as circle ABC – call this circle $\mathcal{C}_{AB}$. Since circles ABC and $\mathcal{C}_{AB}$ are distinct, have the same radius and common chord AB, they must be reflections in AB. That is, $\mathcal{C}_{AB}$ is the reflection of circle ABC in side AB. Let $\mathcal{C}_{BC}$ and $\mathcal{C}_{CA}$ be the reflections of circle ABC in sides BC and CA, respectively. Point P must lie on all three circles $\mathcal{C}_{AB}$, $\mathcal{C}_{BC}$ and $\mathcal{C}_{CA}$. We will show that P is H, the orthocentre of triangle ABC.

Firstly we show that H is on these three circles. Let line AH meet side BC at D and circle ABC again at H'. If we show that H and H' are reflections in side BC, then H lies on circle $\mathcal{C}_{BC}$. By symmetry, H would also lie on $\mathcal{C}_{AB}$ and $\mathcal{C}_{CA}$.

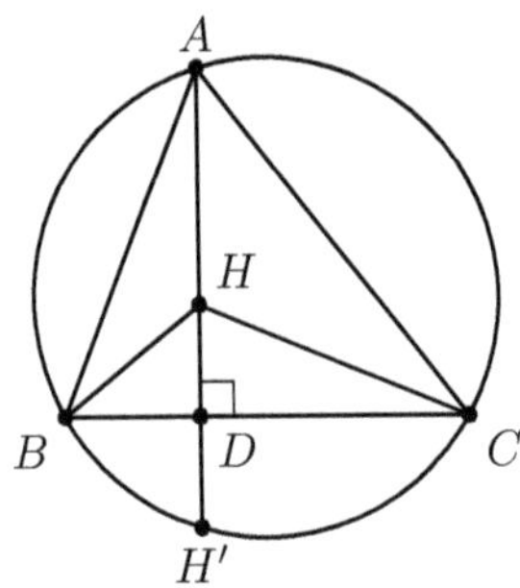

As $ABH'C$ is cyclic, $C\hat{B}H' = C\hat{A}H'$. But AH is perpendicular to BC so $C\hat{A}H' = 90° - B\hat{C}A$. Also, BH is perpendicular to CA so $H\hat{B}C = 90° - B\hat{C}A$. Hence, angles $C\hat{B}H'$ and $H\hat{B}C$ are equal. From this it follows that triangles BHD and $BH'D$ are congruent: they share side BD,

$D\hat{B}H' = H\hat{B}D$ and $H'\hat{D}B = 90° = B\hat{D}H$. Thus, H and H' are indeed reflections in side BC.

Next we show that no other point lies on all three circles. None of $\mathcal{C}_{AB}$, $\mathcal{C}_{BC}$ or $\mathcal{C}_{CA}$ can be circle ABC as P does not lie on circle ABC. In particular, triangle ABC has no right-angles and so H is distinct from the three vertices. Circles $\mathcal{C}_{AB}$, $\mathcal{C}_{CA}$ meet at the two points A and H, circles $\mathcal{C}_{AB}$, $\mathcal{C}_{BC}$ meet at the two points B and H and circles $\mathcal{C}_{BC}$, $\mathcal{C}_{CA}$ meet at the two points C and H. Hence H is the only point lying on all three reflected circles. In particular P must be the orthocentre of triangle ABC and furthermore $S = \{A, B, C, P\}$ satisfies the property in the question.

We will finally show that no further point can be added to S. Suppose we can add the point Q to S. Point Q lies on circle ABC as otherwise Q must be P. Also P must be the orthocentre of triangle QBC so QP is perpendicular to BC – that is, Q lies on line AP. Similarly Q lies on lines BP and CP, so Q must be P, a contradiction.

Problem 3 (Proposed by Dominic Yeo) *A 2019×2019 square grid is made up of 2019^2 unit cells. Each cell is coloured either black or white. A colouring is* balanced *if, within every square subgrid made up of k^2 cells for $1 \le k \le 2019$, the number of black cells differs from the number of white cells by at most one. How many different balanced colourings are there?*

(Two colourings are different if there is at least one cell which is black in exactly one of them.)

Solution by Yuka Machino: First we claim that any balanced colouring does not contain a 3×1 or 1×3 rectangle of one colour. If there is a 3×1 rectangle of one colour, then the three squares to the right and the three squares to the left must be of the opposite colour so that all 2×2 squares are balanced. But then this gives a 3×3 square with six cells of one colour and three cells of another. A similar problem arises if there is a 1×3 rectangle of one colour.

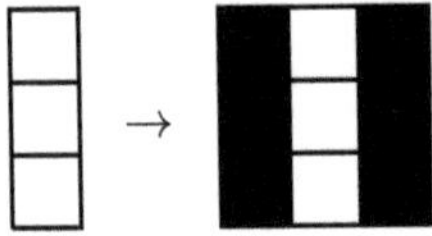

Next we claim that if some row of a balanced colouring alternates black and white, then every row alternates black and white. If this is not the case then there is some row alternating black and white next to a row which has two adjacent squares of the same colour. However, this gives a 2×2 square with three cells of one colour and one of the other. Similarly if some column alternates black and white, then so does every column.

Now we note that if two vertically adjacent cells are of the same colour then the two rows that those square are in are identical and alternate between black and white. Indeed, the two squares to the right and the two squares to the left must both be of the opposite colour. Repeating this gives two identical repeating rows. Call this a 'striped double row'. Similarly if there are two horizontally adjacent cells of the same colour then the two columns they are in form a 'striped double column' – two identical columns which alternate black and white.

Apart from the two chessboard colourings where black and white alternate in every row and column, every colouring must have two adjacent cells which are the same colour. This means that every non-chessboard balanced colouring must have either a striped double row or a striped double column. A colouring cannot have both a striped double row and a striped double column (consider where they meet). Hence the number of balanced colourings is twice the number of balanced colourings with a striped double row plus two (for the two chessboard colourings).

We now consider a balanced colouring with a striped double row. Every row is alternating black and white. This implies that the colouring is determined by the colours of the first column. It also means that, for k even, every $k \times k$ subgrid has the same number of black and white squares since every row in such a subgrid contains an equal number of black and white squares. Called a striped double row 'black' if both rows start with a black square and call a striped double row 'white' if both rows start with a white square.

We now claim that if there is a striped double black row, then the next striped double row must be white (and vice versa). If not, then we have two black striped double rows with alternating rows between. Let there be k rows from the first row of the first striped double black row to the second row of the second striped double black row: k is odd. Consider the $k \times k$ subgrid shown below which consists of the first k squares of each of these rows. As k is odd, if we ignore the first column, then there are an equal number of black and white squares in this subgrid (adjacent columns are inverses of each other). However, the first column contains three more black squares than white squares.

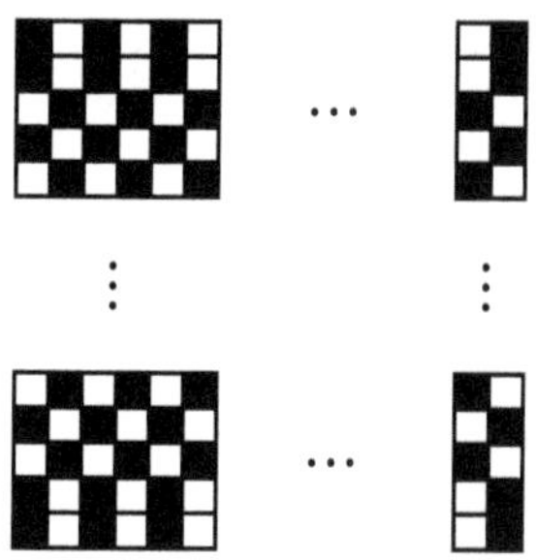

Thus consecutive striped double rows have opposite colours. Hence, if the first striped double row is in the $(2k)^{\text{th}}$ and $(2k+1)^{\text{th}}$ rows, then all other double rows must also be in $(2\ell)^{\text{th}}$ and $(2\ell+1)^{\text{th}}$ rows, while if the first double striped row is in the $(2k-1)^{\text{th}}$ and $(2k)^{\text{th}}$ rows, then all other double rows must also be in $(2\ell-1)^{\text{th}}$ and $(2\ell)^{\text{th}}$ rows. Call the former the 'even' case and the latter the 'odd' case.

To count the even case, we pair up the 2nd and 3rd rows, the 4th and 5th rows, …, and the 2018th and 2019th rows. For each pair we specify whether or not they form a striped double row – as there is at least one striped double row there are $2^{1009}-1$ ways to do this. We also need to specify the colour of the top left square. Specifying all of this determines the grid so there are $2^{1010}-2$ grids in the even case.

To count the odd case, we pair up the 1st and 2nd rows, the 3rd and 4th rows, …, and the 2017th and 2018th rows. For each pair we specify whether or not they form a striped double row and we also specity what colour the top left square of the grid is. This gives $2^{1010} - 2$ grids in the odd case.

If all of the grids in the even case and odd case are balanced grids then the total number of balanced grids is

$$2 \times (2^{2010} - 2 + 2^{2010} - 2) + 2 = 2^{2012} - 6.$$

We will now check that all of the grids in these cases are balanced. We noted above that, for k even, any $k \times k$ subgrid has an equal number of black and white squares. Now consider some $k \times k$ subgrid with k odd. As before, if we ignore the first column, then the number of white and black squares are equal (adjacent columns are inverses of each other). Hence, we just need to check that the number of black and white squares in the first column differ by one. Within the subgrid, consider the topmost striped double row and suppose it is white and consider the bottommost striped black double row. As consecutive striped double rows alternate in colour and between striped double rows the squares alternate in colour, there an equal number of black and white squares from this top white double to the bottom black double. Ignoring this range leaves some alternating squares possibly with one double striped row remaining. The number of white and black squares in this remainder differ by one.

Hence, the number of balanced grids is $2^{2012} - 6$.

Problem 4 (Proposed by Jeremy King) *A sequence $b_1, b_2, b_3, \ldots$ of nonzero real numbers has the property that*

$$b_{n+2} = \frac{b_{n+1}^2 - 1}{b_n}$$

for all positive integers n.

Suppose that $b_1 = 1$ and $b_2 = k$ where $1 < k < 2$. Show that there is some constant B, depending on k, such that $-B \leq b_n \leq B$ for all n. Also show that, for some $1 < k < 2$, there is a value of n such that $b_n > 2020$.

This question was found to be very hard. All solutions found an expression for b_n in terms of n and then used this expression to get the bounds in the question. Most successful students showed that the sequence (b_n) satisfied the recurrence relation $b_{n+2} = kb_{n+1} - b_n$. Many did this by inspecting the first few terms of the sequence and then proving it by induction, although it is possible to deduce the recurrence by rearranging the equation given.

Solution 1 to first part by Soren Choi: We may rearrange the equation given to $b_{n+1}^2 - b_n b_{n+2} = 1$. In particular, $b_{n+1}^2 - b_n b_{n+2} = 1 = b_n^2 - b_{n-1} b_{n+1}$ and so $b_{n+1}(b_{n+1} + b_{n-1}) = b_n(b_{n+2} + b_n)$. Hence,

$$\frac{b_{n+2} + b_n}{b_{n+1}} = \frac{b_{n+1} + b_{n-1}}{b_n}.$$

The right-hand side of this equation is the same as the left but with n replaced by $n - 1$. Therefore

$$\frac{b_{n+2} + b_n}{b_{n+1}} = c,$$

where c is a constant that does not depend upon n. Now $b_3 = k^2 - 1$, so $c = (b_3 + b_1)/b_2 = k$ and hence $b_{n+2} + b_n = kb_{n+1}$.

The recurrence $b_{n+2} = kb_{n+1} - b_n$ is a second order linear recurrence with associated quadratic $x^2 - kx + 1 = 0$. If this quadratic has roots r_+ and r_-, then $b_n = \alpha r_+^n + \beta r_-^n$ where α and β are (complex) constants not depending upon n. Solving the quadratic gives

$$r_+ = \tfrac{k}{2} + i\sqrt{1 - k^2/4} \quad \text{and} \quad r_- = \tfrac{k}{2} - i\sqrt{1 - k^2/4}.$$

Now r_+ and r_- both have magnitude 1, so $|\alpha r_+^n| = |\alpha|$ and $|\beta r_-^n| = |\beta|$. Thus, $|b_n| \leqslant |\alpha| + |\beta|$ and so the constant $B = |\alpha| + |\beta|$ satisfies $-B \leqslant b_n \leqslant B$ for all n.

Solution 2 to first part by Benedict Randall Shaw: We first note the trigonometric identity $\sin(x+y)\sin(x-y) = \sin^2(x) - \sin^2(y)$ which can be proved by expanding out the left-hand side using the formulae for the sine of a sum/difference of two angles.

Let $c = \arccos(k/2)$ – we claim that

$$b_n = \frac{\sin(cn)}{\sin(c)}.$$

We use a_n to denote the right-hand side – we wish to show $a_n = b_n$ for all n. First note that $a_1 = \frac{\sin(c)}{\sin(c)} = 1$ and $a_2 = \frac{\sin(2c)}{\sin(c)} = 2\cos(c) = k$, so $a_1 = b_1$ and $a_2 = b_2$. We just need to show that the sequence (a_n) satisfies the same relation as (b_n). Now,

$$\begin{aligned} a_{n+2}a_n &= \tfrac{1}{\sin^2(c)} \sin(c(n+2))\sin(cn) \\ &= \tfrac{1}{\sin^2(c)}[\sin^2(c(n+1)) - \sin^2(c)] \\ &= a_{n+1}^2 - 1, \end{aligned}$$

so we indeed have $b_n = a_n$ for all n. Sine takes values between -1 and 1, so

$$-\tfrac{1}{|\sin(c)|} \leqslant b_n \leqslant \tfrac{1}{|\sin(c)|},$$

holds for all n.

Solution to second part by Matthew Stevens: As in solution 1, we find $b_{n+2} = kb_{n+1} - b_n$ for all n. For motivation, we consider the case where $k = 2$. When $k = 2$, we can show, by induction, that $b_n = n$ for all n: this is true when $n = 1$ and $n = 2$ and furthermore, if it is true for n and $n+1$, then $b_{n+2} = 2(n+1) - n = n+2$.

Instead, we consider $k = 2 - \varepsilon$ and claim that $b_n = P_n(\varepsilon)$ where P_n is a polynomial with constant term n. Indeed, this is true when $n = 1$ and when $n = 2$. Suppose it is true for n and $n+1$. Then

$$\begin{aligned} b_{n+2} &= (2-\varepsilon)b_{n+1} - b_n \\ &= (2-\varepsilon)P_{n+1}(\varepsilon) - P_n(\varepsilon). \end{aligned}$$

Now $P_{n+2}(x) = (2-x)P_{n+1}(x) - P_n(x)$ is a polynomial whose constant term is $2(n+1) - n = n+2$ which completes the induction. The polynomial $P_{2021}(x)$ has constant term 2021, so, if ε is positive but very close to zero, then $P_{2021}(\varepsilon) > 2020$. Hence, for $k = 2 - \varepsilon$ we would have $b_{2021} > 2020$.

2.14.2 Participation

Participation in the British Mathematical Olympiad Round 2

Year of competition	2015–16	2016–17	2017–18	2018–19	2019–20
Centres with invitees	78	74	71	72	74
Centres returning scripts	117	103	120	113	126
Invitees	114	119	93	108	105
Scripts returned	221	243	220	237	245

Percentage change from previous year

Year of competition	2016–17	2017–18	2018–19	2019–20
Centres with invitees	-5.1%	-4.1%	+1.4%	+2.7%
Centres returning scripts	-12.0%	+16.5%	-5.8%	+11.5%
Invitees	+4.4%	-21.8%	+16.1%	-2.7%
Scripts returned	+10.0%	-9.5%	+7.7%	+3.37%

2.14.3 Student performance

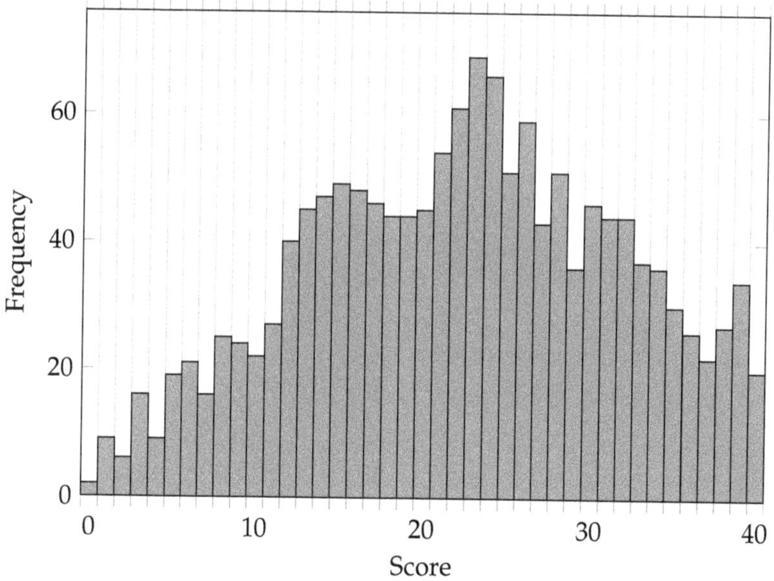

Mean scores and award boundaries

Based on the results of this competition, with allowance for age and experience, 20 or so students are invited to a training camp at Trinity College, Cambridge.

Year of competition	2015–16	2016–17	2017–18	2018–19	2010–20
Certificate of Distinction	–	15	19	15	15
Mean score	5.9	10.7	11.8	10.7	10.2

2.15 Mathematical Olympiad for Girls

2.15.1 Questions, solutions and markers' comments

United Kingdom
Mathematics Trust

Mathematical Olympiad for Girls

Tuesday 8 October 2019

Instructions

1. Do not turn over the page until told to do so.
2. Time allowed: $2\frac{1}{2}$ hours.
3. Each question carries 10 marks. Full marks require clearly written solutions — not just answers — including complete proofs of any assertions you may make.
   Marks awarded will depend on the clarity of your mathematical presentation. Work in rough first, and then write up your best attempt.
4. Partial marks may be awarded for good ideas, so try to hand in everything that documents your thinking on the problem — the more clearly written the better.
   However, one complete solution will gain more credit than several unfinished attempts.
5. Earlier questions tend to be easier. Some questions have multiple parts. Often earlier parts introduce results or ideas useful in solving later parts of the problem.
6. The use of rulers and compasses is allowed, but calculators and protractors are forbidden.
7. Start each question on a fresh sheet of paper. Write on one side of the paper only.
   On each sheet of working write the number of the question in the top left-hand corner and your initials and the centre number (but NOT your name or school) in the top right-hand corner.
8. Complete the cover sheet provided and attach it to the front of your script, followed by your solutions in question number order.
9. Staple all the pages neatly together in the top left hand corner.
10. To accommodate candidates sitting in other time zones, please do not discuss the paper on the internet until 08:00 BST on Wednesday 9 October.

Enquiries about the Mathematical Olympiad for Girls should be sent to:

1. At Mathsland Animal Shelter there are only cats and dogs. Unfortunately, one day 60 of the animals managed to escape. Once a volunteer had realised, they counted the remaining animals. They noted that half of the cats and a third of the dogs had escaped.

(a) (i) If the number of cats before the escape was C and the number of dogs before the escape was D, write down an equation linking C and D.

(ii) If the total number of animals before the escape was T, write down an equation linking C, D and T.

(4 marks)

(b) Given that more cats than dogs escaped, find the largest possible value of T. You must justify why the value you have found is the largest. (6 marks)

SOLUTION

COMMENTARY

This is similar to standard simultaneous equations questions you have probably met at school. The difference is that, once you have written down the two equations, you will see that you don't have enough information to uniquely determine the values of C, D and T.

You may want to start by trying to find the values of C and D for various values of T. Can you always find a solution that works? Remember that the question places some constraints on possible values of C and D.

To justify that the value you have found for T is the largest possible, you must show that this value can be achieved (by showing an example of C and D in that case) and also that any value larger that is cannot be achieved.

(a) The total number of escaped animals is 60, and the total number of animals before the escape is T.

(i) $\frac{1}{2}C + \frac{1}{3}D = 60$

(ii) $C + D = T$

(b) From the first equation, $D = 180 - \frac{3}{2}C$. Substituting into the second equation,

$$T = C + (180 - \tfrac{3}{2}C) = 180 - \tfrac{1}{2}C.$$

Therefore T is the largest possible when C is the smallest possible. Since more half of the 60 escaped animals were cats, $\frac{1}{2}C \geq 31$, so $T \leq 180 - 31 = 149$. This value is achieved with $C = 62$ and $D = 87$.

Hence the largest possible value of T is 149.

COMMENTARY

An alternative approach for part (b) is to solve the equations for C and D in terms of T.

Rewrite the first equation as $3C + 2D = 360$. Multiplying the second equation by 2 and subtracting from the first gives $C = 360 - 2T$. Substituting back then gives $D = 3T - 360$.

You now need to consider the constraints on C and D. First, both need to be non-negative, so $T \leq 180$ and $T \geq 120$. The condition that more cats and dogs escaped means that $\frac{1}{2}C > \frac{1}{3}D$ or, equivalently, $3C > 2D$. Substituting the expressions for C and D in terms of T gives:

$$3(360 - 2T) > 2(3T - 360).$$

Rearranging this inequality gives $T < 150$ so, since T is an integer, the largest possible value of T is 149. This value is achieved when $C = 62$ and $D = 87$, which satisfy the conditions of the problem.

2. Beth has a black counter and Wendy has a white counter. Beth and Wendy move their counters on the two boards below according to the starting positions and rules given. They always move their counters at the same time.

(a) 

At each turn, each player moves their counter either one square to the left or one square to the right. Prove that the black and white counters can never be in the same square at the same time.

HINT You may find it helpful to refer to the colours of the squares on the board in your explanation. (3 marks)

(b) At each turn, each player moves their counter to a triangular cell which shares one edge with the cell that their counter is currently in. Can their counters ever be in the same cell at the same time? (7 marks)

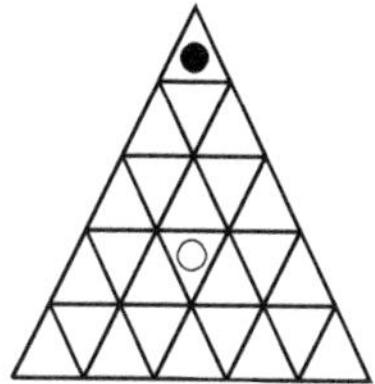

HINT If you think the two counters can never be in the same cell at the same time, you should give an argument that they cannot be in the same cell at the same time which works no matter which sequence of moves Beth and Wendy do. If you think the two counters can be in the same cell at the same time, you should give an example of a sequence of moves after which they are in the same cell at the same time.

SOLUTION

COMMENTARY

In a problem like this, it is tempting to try to think of all possible ways that the two counters can move and find a "worst possible" or "best possible" sequence of moves. Another strategy would be to consider where the counters were the move before they were in the same square and then "work backwards" to the starting position. It turns out that, in many cases, both of these strategies are either impractical (because there are too many possible sequence of moves) or impossible (because, like here, the counters can keep moving forever without meeting).

Instead, you need to think of some property of the game that means that the counters can never be in the same square. The hint in part (a) suggests thinking about how the colour of each counter's square changes with each move.

Once you have seen that considering the colours of the cells is useful in part (a), you may want to try a similar strategy in part (b). You need to decide how to colour the triangular cells so that each move changes the colour of the counter's cell. You also need to show that, with your colouring, the black and the white counter start on different colours.

(a) Since the colours of the squares on the board alternate, each move (one square to the left or one square to the right) changes the colour of the square that the counter is in. The two counters start on different colours so, since each one changes the colour at each turn, they will always be on different colours. This means that they can never be in the same square at the same time.

(b) Colour the triangular cells black and white as shown in the diagram.

Then two cells which share an edge are different colours. Hence each move changes the colour of the counter's cell.

As can be seen in the diagram, the black and the white counter start on different colours. Therefore they will always be on different colours, and so cannot be in the same cell at the same time.

3. (a) Seth wants to know how many positive whole numbers from one to one hundred are divisible by two or five. He thinks that the answer is 70 because there are fifty multiples of two and twenty multiples of five from one to one hundred. Explain why his answer is too large. (2 marks)

(b) Consider the list of 1800 fractions

$$\frac{1}{1800}, \frac{2}{1800}, \ldots, \frac{1799}{1800}, \frac{1800}{1800}.$$

How many are *not* in simplest form? Explain your reasoning. (8 marks)

[Note: The fraction $\frac{900}{1800}$ is not in simplest form because it can be simplified to $\frac{1}{2}$.]

Solution

Commentary

The main part of the question (part (b)) asks how many of the fractions can be simplified. How can you tell when a fraction can be simplified? Part (a) should help you avoid a common trap when counting multiples.

It may help to start by writing out some multiples of 2 and 5. Which numbers will be listed more than once? This tells you not only why 70 is too large, but also helps you find the correct answer to Seth's question.

In part (b) you need to think carefully what exactly you want to count. For example, 1800 is divisible by both 3 and 9, but do you need to count the multiples of 3 and 9 separately?

You also need to extend the reasoning from part (a) to avoid Seth's trap. You may find it helpful to use a Venn diagram to represent the number of multiples.

(a) If we list all 50 multiples of 2 and all 20 multiples of 5, all the multiples of 10 will appear in both lists. Since some numbers are counted twice, the real answer is smaller than 70.

(b) A fraction will not be in simplest form when the numerator shares at least one factor with 1800. Since $1800 = 2^3 \times 3^2 \times 5^2$, we need to count how many of the numbers from 1 to 1800 are divisible by 2, 3 or 5.

Imagine making three separate lists: one containing the multiples of 2, one containing the multiples of 3 and one containing the multiples of 5. There are $\frac{1800}{2} = 900$ numbers in the first list, $\frac{1800}{3} = 600$ in the second and $\frac{1800}{5} = 360$ in the third, a total of 1860 numbers.

The multiples of 6, 10 and 15 appear twice, so we need to take away $\frac{1800}{6} + \frac{1800}{10} + \frac{1800}{15} = 600$.

However, the multiples of 30 have now been taken away three times. But they appear three times in the original three lists, so they should have only been taken away twice. Hence we need to add back $\frac{1800}{30} = 60$.

The required total is therefore 1860 – 600 + 60 = 1320. Thus 1320 of the fractions are not in simplest form.

COMMENTARY

Try drawing a Venn diagram to show the number of multiples of 2, 3 and 5. Does that make our calculation clearer?

This idea can be extended to count the total number of elements in more than three overlapping sets, resulting in what is know as the inclusion-exclusion principle.

4. The diagram shows a rectangle placed inside a quarter circle of radius 1, such that its vertices all lie on the perimeter of the quarter circle and one vertex coincides with the centre of the (whole) circle.

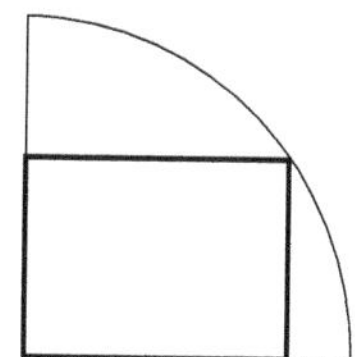

Let the perimeter of such a rectangle be P.

(a) Show that $P = 3$ is impossible. (4 marks)

(b) Find the largest possible value of P. You must fully justify why the value that you find is the largest. (4 marks)

Instead a rectangle is placed inside a whole circle of radius 1, such that its vertices all lie on the circumference of the circle.

(c) If the perimeter of the rectangle is as large as possible, show that the rectangle *must* be a square and calculate its perimeter. (2 marks)

Solution

Commentary

Call the sides of the rectangle x and y. It seems reasonable to start by writing some equations connecting x, y and P. Looking at parts (a) and (b) together, it sounds like these equations will only have a solution for some values of P.

One vertex of the rectangle is the centre of the circle and the opposite vertex is on the circumference, so you can use Pythagoras's Theorem to link x and y with the radius of the circle. This means that the equation you get will be quadratic, so you can expect to use the discriminant to determine whether it has any solutions.

In part (b) you need to show two things: that the value of P cannot be larger than the one you found, but also that there is a rectangle with this value of P.

In part (c), you could start again by writing equations connecting the sides and the perimeter of the second rectangle. However, if you split the circle into quarters, then each quarter has a rectangle inscribed in it in the same way as in part (b). You can therefore use the results you fond in part (b) about the largest possible value of the perimeter.

In the solution below we start by deriving the quadratic equation which will be used in all three parts.

Let x and y be the sides of the rectangle. The diagonal of the rectangle is a radius of the circle, so $x^2 + y^2 = 1$. The perimeter of the rectangle is $P = 2x + 2y$. Substituting $y = \dfrac{P - 2x}{2}$ from the second equation into the first gives

$$x^2 + \left(\frac{P - 2x}{2}\right)^2 = 1,$$

which is equivalent to

$$8x^2 - (4P)x + (P^2 - 4) = 0.$$

(a) When $P = 3$ this quadratic equation becomes $8x^2 - 12x + 5 = 0$. The discriminant is $12^2 - 4 \times 8 \times 5 = -16 < 0$ so there are no solutions for x. It is therefore not possible that $P = 3$.

(b) We now look for values of P for which the quadratic equation $8x^2 - (4P)x + (P^2 - 4) = 0$ has a solution. The discriminant needs to be non-negative, so we need

$$(4P)^2 - 32(P^2 - 4) \geq 0.$$

This is equivalent to $16P^2 \leq 128$ and, since $P > 0$, we must have $P \leq 2\sqrt{2}$. For this value of P, solving the quadratic equation gives $x = \frac{\sqrt{2}}{2}$ and, substituting back into $2x + 2y = 2\sqrt{2}$, $y = \frac{\sqrt{2}}{2}$. Thus it is possible that $P = 2\sqrt{2}$ and this is the largest possible value of P.

(c) Let the sides of the rectangle be $2x$ and $2y$. The diagonal of the rectangle is a diameter of the circle, so $x^2 + y^2 = 1$. (Note that this is the same relationship between x and y as in part (b).) The perimeter of the rectangle is $4x + 4y = 2(2x + 2y)$, which is twice the perimeter from part (b).

We know from part (b) that the largest possible value of $2x + 2y$ is $2\sqrt{2}$, and that this value only occurs when $x = y = \frac{\sqrt{2}}{2}$. Therefore the largest possible value of our perimeter is $4\sqrt{2}$ and it occurs when the sides of the rectangle are equal, i.e. when it is a square.

5. Let n be a positive integer. Tracy writes a list of 10 whole numbers between 1 and n (inclusive). Each number in the list is either equal to, one less than, or one more than the number before it.

For example, when $n = 7$:

Her list could be 5, 5, 6, 7, 6, 6, 5, 5, 6, 6 or 4, 4, 3, 2, 1, 1, 1, 1, 1, 1.

Her list could *not* be 1, 3, 3, 4, 5, 5, 6, 7, 7, 7 or 5, 6, 7, 8, 7, 6, 5, 5, 5, 5.

(a) Suppose that $n = 3$. Stacey forms a list by copying Tracy's list, except that whenever Tracy writes a 1, Stacey writes a 3, and whenever Tracy writes a 3, Stacey writes a 1.

(i) Which lists could Tracy write that would cause her list to be the same as Stacey's?

(ii) Explain why Tracy can write as many lists that start 2, 2, 1 as start 2, 2, 3.

(3 marks)

(b) For which n between 1 and 10 (inclusive) is the number of lists that Tracy could write odd? (7 marks)

SOLUTION

COMMENTARY

There seems to be a lot going on in this question, so it is probably a good idea to start by writing out some lists. First make sure that you understand the rules. For $n = 3$, can you write some of Tracy's lists and the corresponding Stacey's lists? What makes the lists the same? What makes them different?

Part (a)(ii) asks you to consider lists of two different types (those starting 2, 2, 1 and those starting 2, 2, 3) and to show that there is the same number of each type. A useful way of showing that two sets contain the same number of elements is to try and pair them up. Can you see how Stacey can help you? You may well find that thinking in terms of Tracy's and Stacey's lists is the easiest way to write an explanation.

In part (b), you will need to adapt Stacey's rule slightly. Don't forget to check and explain why Stacey's new rule always generates a valid list (where each number is either equal to, one less or one more than the number before it). If the total number of Tracy's lists is odd, what does that tell you about pairing them up with Stacey's lists?

(a) (i) If Tracy writes a 3 at any point in her list, then Stacey will write a 1 at that point and so the lists will be different. Also, is Tracy writes a 1 at any point in her list, Stacey will write a 3 at that point and so the lists will be different. So, the only list that Tracy could possibly write that would cause Stacey to write the same list is the list where all ten entries are 2s. Indeed, this list does cause Stacey to have the same list as Tracy.

(ii) Note that whenever Tracy writes down a valid list of numbers, the list Stacey writes down is also a valid list of numbers. For any list L that Tracy can write, let $S(L)$ be the list that this causes Stacey to write down.

Suppose Tracy writes a list L that starts 2,2,1. Then $S(L)$ will start 2,2,3. We can get back to L from $S(L)$ by replacing all of the 1s in $S(L)$ with 3s and all of the 3s in $S(L)$ with 1s. In symbols, this says:

$$S(S(L)) = L.$$

So we can pair up each list L that Tracy could write beginning 2,2,1 with the list $S(L)$ beginning 2,2,3 that Stacey writes. Since this gives every list beginning 2,2,1 a unique partner beginning 2,2,3 (and vice versa), there must be the same number of lists beginning 2,2,1 as begin 2,2,3.

(b) Suppose for each n that whenever Tracy writes a list, Stacey copies Tracy's list except that whenever Tracy writes k, Stacey writes $n+1-k$. If Tracy writes a list L, write $S(L)$ again for the list that this causes Stacey to write.

Suppose Tracy writes a list which causes Stacey to write down the same list, i.e. $L = S(L)$. If an entry in Tracy's list is k, then we must have $k = n+1-k$, so $k = \frac{n+1}{2}$. Therefore, if n is odd there is one list where Stacey and Tracy write down the same list, namely

$$\frac{n+1}{2}, \frac{n+1}{2}, \frac{n+1}{2}, \frac{n+1}{2}, \frac{n+1}{2}, \frac{n+1}{2}, \frac{n+1}{2}, \frac{n+1}{2}, \frac{n+1}{2}, \frac{n+1}{2}.$$

If n is even, then $\frac{n+1}{2}$ is not a whole number so there are no lists that cause Stacey and Tracy to write down the same list.

Now, if Tracy writes the list L and this causes Stacey to write the list $S(L)$, we can recover L from $S(L)$ by replacing each entry of $n+1-k$ with k, for $1 \leq k \leq n$. Since $n+1-(n+1-k) = k$, this is the same as doing Stacey's operation a second time, so

$$S(S(L)) = L.$$

So, all of the possible lists can be broken up into pairs of lists, where we pair the list L with its partner $S(L)$. If no lists are paired with themselves, this splits all of the possible lists into pairs, so there must be an even number of possible lists. Therefore, in the case n is even, there is an even number of possible lists. In the case n is odd, there is one list M that is paired with itself. So, the number of possible lists except for M is even, and therefore the total number of possible lists including M is one more than an even number. So, the total number of possible lists is odd whenever n is odd.

The possible values of n for which Stacey can write an odd number of lists are 1, 3, 5, 7 and 9.

General comments from markers

Students engaged very well with this paper and the markers enjoyed reading many imaginative and well explained solutions. A vast majority of candidates attempted all the questions and were often able to score at least one or two marks even on the later ones. Over two-thirds of the candidates solved at least one question fully. In last year's report we commented on the need to consider part (a) of a question as a hint for doing part (b). This advice was followed well this year, particularly in Questions 2 and 3 where we saw many good attempts to use a colouring argument, or to find overlaps between sets of multiples. Even in Question 5, a record number of candidates made progress in part (b) by trying to pair up the lists in some way, as suggested by part (a). There were several common ways to lose marks. Questions 1 and 4 both asked for a maximum possible value of a quantity; often the students found the required value but did not justify why it is a maximum, or show that it can in fact be achieved within the constraints of the question. In Questions 1, 2 and 5 it was sensible to start by trying some examples (possible combinations of cats and dogs; first few moves of the counters; continuing Tracy's lists). However, those examples can only suggest a pattern or rule that then needs to be explained and justified. It was somewhat disappointing to see algebraic mistakes in Questions 1 and 4 when dealing with fractions and square roots. Overall, however, the quality of candidates was very high. We were particularly impressed by solutions to Question 3, which produced the largest number of well-explained, structured arguments. The 2019 Mathematical Olympiad for Girls attracted 1848 entries. The scripts were marked on 19th and 20th October in Cambridge by a team of Amelia Rout, Amit Shah, Anđela Šarković, Andreas Stavrou, Andrew Carlotti, Arij Asad, Abhilasha Aggarwal, Emily Beatty, Eve Pound, James Cranch, Jeremy King, Jerome Watson, Joseph Myers, Kasia Warburton, Lulu Beatson, Martin Orr, Melissa Quail, MT Fyfe, Oscar Hidalgo, Philip Coggins, Phillip Beckett, Robin Bhattacharyya, Stephen Tate, Vesna Kadelburg and Zarko Randjelovic.

2.15.2 Participation

Participation in the Mathematical Olympiad for Girls

Year of competition	2016	2017	2018	2019
Centres placing orders	364	380	310	400
Centres returning scripts	344	373	302	388
Papers ordered	2,821	1,968	1,622	2,089
Scripts returned	2,129	1,780	1,475	1,860

Percentage change from previous year

Year of competition	2017	2018	2019
Centres placing orders	+4.4%	-18.4%	+29.0%
Centres returning scripts	+8.4%	-19.0%	+28.48%
Papers ordered	-30.2%	-17.6%	+24.79%
Scripts returned	-16.4%	-17.1%	+26.1%

2.15.3 Student performance

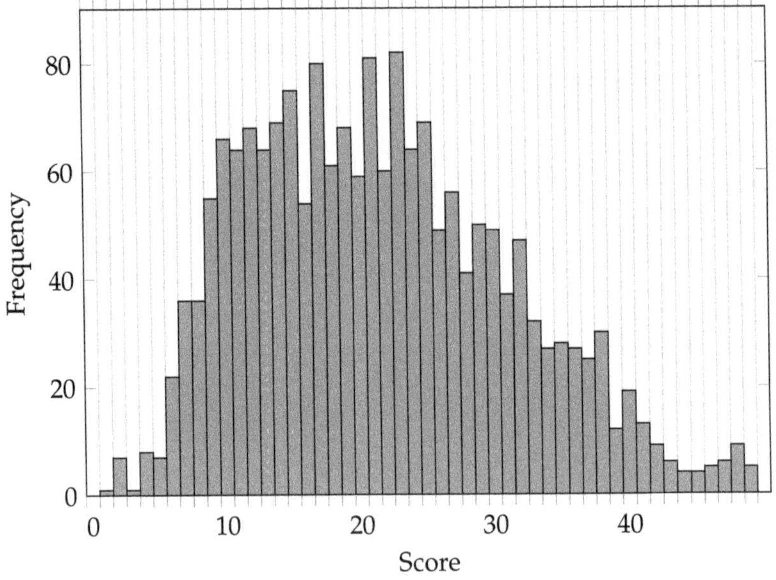

Mean scores and award boundaries

The table below shows the minimum score needed to obtain each award, and the mean score, for each of the past four years of the competition.

Year of competition	2016	2017	2018	2019
Prize	30	40	45	46
Certificate of Distinction	12	14	27	28
Honourable Mention	10	-	-	-
BMO1 invitation	26	31	40	41
Maclaurin invitation	21	21	30	-
Hamilton invitation	15	15	20	-
Cayley invitation	11	24	15	-
Mean score	6.3	9.5	19.6	21.7

Chapter 3

Team competitions and resources

3.1 Primary Team Maths Resources

We have several sets of Primary Team Maths Resources. These are produced with the aim to facilitate secondary schools in running maths events for feeder schools. Schools may choose to use the materials in other ways. For example, a primary school may use the materials to run a competition for their own pupils, or a secondary school may use the materials as an end of term activity for their younger students.
The Primary Team Maths Resources may be downloaded by registered UKMT Centres, free of charge, from our website.

3.2 Team Maths Challenge

The Team Challenge promotes mathematical dexterity, teamwork and communication skills. It also gives students the opportunity to compete against students from other schools in their region. The competition is the successor of Enterprising Mathematics UK which was run in conjunction with the International Mathematical Olympiad in 2002.
A team in the Team Maths Challenge consists of four students from Years 8–9 (England and Wales), S1–S2 (Scotland) or Years 9–10 (Northern Ireland) with no more than two students from the upper year group.

Regional Finals take place in dozens of venues across the UK during February–April. Each team travels with an accompanying teacher and competes against teams from other schools in the area in a series of four rounds. Unfortunately, The 2020 competition was cancelled midway through the regional events.

Group Round The team has 45 minutes to answer 10 questions. It is for the team to decide on a strategy to tackle the problems.

Crossnumber The time allowed for this round is 45 minutes. Teams divide into pairs and each pair works independently on either the ACROSS clues or the DOWN clues. The pairs may only communicate with each other by entering digits into the grid or by asking the other pair to try to solve a particular clue.

Shuttle Teams split into pairs. One pair is given questions 1 and 3; the other pair is given questions 2 and 4. The first pair works on question 1 and passes the answer to the other pair who use it to help answer question 2. This continues with the second pair passing the answer back to the first pair, and so on, until a full set of answers is provided for marking. Bonus points are awarded to teams which present a correct set of answers before the six-minute whistle, then the other teams have another two minutes to finish. Four such shuttles are attempted in this round.

Relay Each team is split into two pairs, with each pair seated at a different desk away from the other pair and their supervising teacher. One member of Pair A collects question A1 from the supervising teacher and returns to their partner to answer the question together. When the pair is certain that they have answered the question, the runner returns to the front to submit their answer to the supervising teacher. If it is correct, the runner is given question B1 to give to the other pair (Pair B) from their team. If it is incorrect, Pair A then has a second (and final) attempt at answering the question, then the runner returns to the front to receive question B1 to deliver to pair B. The runner then returns, empty handed, to their partner. Pair B answers question B1 and a runner from this pair takes the answer to the supervising teacher, as above, then takes question A2 to Pair A. Pair A answers question A2, their runner returns it to the front and collects question B2 for the other pair, and so on until all questions are answered or time runs out. Thus Pair A answers only A questions and Pair B answers only B questions. Only one pair from a team should be working on a question at any time and each pair must work independently of the other.

High-scoring teams from the Regional Finals are invited to compete against each other at the National Final held in London in June. All participants receive a certificate and goody bag, and the winners are presented with prizes and the Team Maths Challenge trophy. The rounds in the National Final are the same as the Regional Finals, except for the following.

Poster Competition This is an additional round where teams are given 50 minutes to design a poster on a given topic.

Group Circus This replaces the Group Round and includes more practical materials for use in solving the problems. Unfortunately, the 2020 National Final did not take place this year at all.

3.2.1 Regional Final materials

Group Round

Group Round

Team Maths Challenge
2019
Regional Final

Instructions

- Your team will have 45 minutes to answer 10 questions. Each team will have the same questions.
- Each question is worth a total of 6 marks. However, some questions are easier than others!
- Do not spend too long on any one question without sharing it with the rest of the team.
- You will have to decide your team's strategy for this group competition.
- There is only one response sheet per team.
- Remember to finalise your answers and write them on the response sheet before the end of the round.

GROUP ROUND

TEAM MATHS CHALLENGE
2019
REGIONAL FINAL

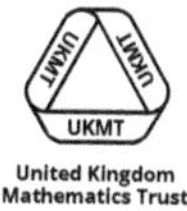

QUESTION 1

The basic State Pension used to be £98.70 per week.

This amount was increased by $2\frac{1}{2}\%$ and rounded up to the nearest penny.

What is the new basic State Pension in pounds, per week?

[6 marks]

GROUP ROUND

TEAM MATHS CHALLENGE
2019
REGIONAL FINAL

QUESTION 2

Naomi had one £2 coin, two £1 coins, three 50p coins, four 20p coins, five 10p coins, six 5p coins, seven 2p coins and eight 1p coins in her purse.

(a) Firstly Naomi bought some food, costing £5.39.

As she was in a rush to pick up her children from school, she paid the exact amount with the least number of coins possible.

How many coins did she have left in her purse? [3 marks]

(b) Next, Naomi went straight to school, arriving ten minutes before she needed to pick up her children.

Whilst waiting, she bought them some sweets, costing 99p.

Having a bit of time, she decided to pay this exact amount with as many coins from her purse as possible.

How many coins did she have left in her purse?

[3 marks]

GROUP ROUND

TEAM MATHS CHALLENGE
2019
REGIONAL FINAL

QUESTION 3

(a) $14 \times (d + e + f) =$ 'def' where d, e and f represent different digits and 'def' is a three-digit number.

What is the value of 'def'? [3 marks]

(b) $19 \times (a + b + b) =$ 'abb' where a and b represent different digits and 'abb' is a three-digit number.

What is the maximum possible value of 'abb'?

[3 marks]

Group Round

Team Maths Challenge
2019
Regional Final

Question 4

(a) Jo took 10% *more* time to run the second half of a marathon than she took to run the first half.

It took her 1 hour 28 minutes to run the second half of the marathon.

How long did Jo take to run the marathon, in hours and minutes? [3 marks]

(b) Usain Bolt ran the 200m in 19.19 seconds.

It took him 10% *less* time to run the second 100m than it took him to run the first 100m.

How long in seconds did it take him to run the first 100m? [3 marks]

GROUP ROUND

TEAM MATHS CHALLENGE
2019
REGIONAL FINAL

QUESTION 5

In the Olympics, the order that a country is placed in the medal table is based firstly on the number of Gold medals they win, then the number of Silver medals and finally the number of Bronze medals.

A possible outcome is:

	Position	Gold	Silver	Bronze
Kenya	15th	2	1	0
Jamaica	16th	1	9	3
Croatia	17th	1	8	22

With three medals available the positions of Great Britain and China were as follows:

	Position	Gold	Silver	Bronze
China	2nd	27	32	48
Great Britain	3rd	26	31	47

China won no medals of the last three available.

In how many different ways could Great Britain finish above China in the medal table?

[6 marks]

GROUP ROUND

TEAM MATHS CHALLENGE
2019
REGIONAL FINAL

QUESTION 6

A digit is to be placed in each of the nine blank cells in the 3×3 grid shown below.

Every pair of neighbouring digits vertically down or horizontally to the right forms a two-digit number.

If all of the two-digit numbers are squares, what is the sum of the numbers that must be placed in the four corners?

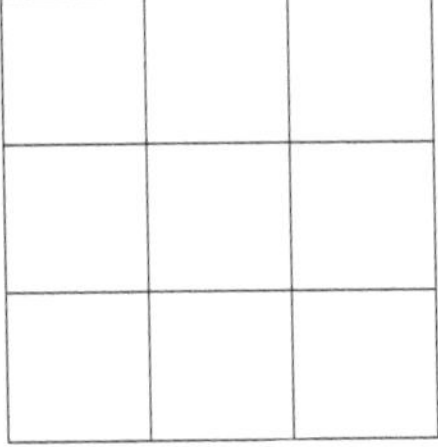

[6 marks]

GROUP ROUND

TEAM MATHS CHALLENGE
2019
REGIONAL FINAL

QUESTION 7

The reverse number of the year 2019 is 9102.

The sum of these two numbers is 11121.

(a) After 2019, what is the first year that when added to its reverse number equals 11121? [3 marks]

(b) After 2019, what is the second year that when added to its reverse number equals 11121? [3 marks]

GROUP ROUND

TEAM MATHS CHALLENGE
2019
REGIONAL FINAL

QUESTION 8

The sum of the page numbers, in a book, is less than 2019.

The pages are numbered in order, from 1 to n.

What is the largest possible value of n?

[6 marks]

GROUP ROUND

TEAM MATHS CHALLENGE
2019
REGIONAL FINAL

QUESTION 9

A regular pentagon and a square share a common side, as shown.

Another regular polygon shares a common side with the pentagon and a common side with the square. Part of this regular polygon is shown.

How many sides does this regular polygon have?

NOTE: The external angle of a regular polygon with n sides is $\frac{360}{n}$ $^\circ$.

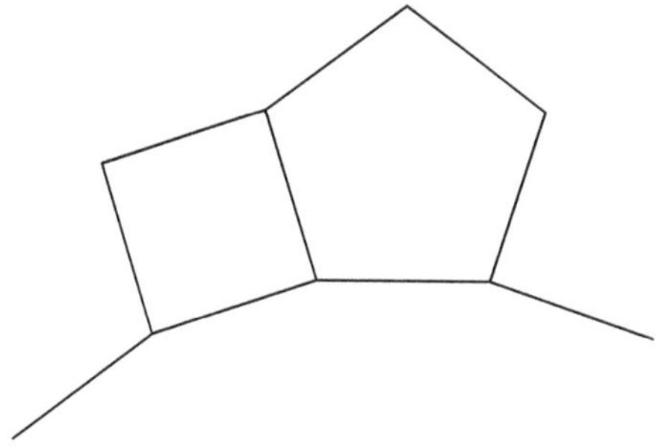

[6 marks]

Group Round

Team Maths Challenge
2019
Regional Final

Question 10

The letters A, B, C, D, E, F, G and H each represent different primes.

(a) $A + B = 14$

$C - D = 17$

What is the value of $D \times C - B \times A$? [3 marks]

(b) $E + F + G = H$

What is the smallest possible value of H? [3 marks]

GROUP ROUND

ANSWERS

1.

£101.17

2.

(a) 26 (b) 5

3.

(a) 126 (b) 399

4.

(a) 2 hours 48 minutes (b) 10.1 seconds

5.

5

6.

29

7.

(a) 2109 (b) 3018

8.

63

9.

20

10.

(a) 5 (b) 29

On the RESPONSE SHEET: Circle the mark awarded for each question and cross out the others.

Crossnumber

CROSSNUMBER

TEAM MATHS CHALLENGE
2019
REGIONAL FINAL

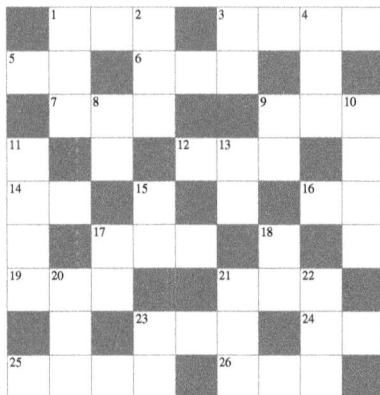

	1		2		3		4	
5			6					
	7	8				9		10
11				12	13			
14			15				16	
		17				18		
19	20				21		22	
			23				24	
25					26			

ACROSS

1. The remainder when 11 DOWN is divided by 19 ACROSS (3)

3. The mean of 25 ACROSS and 10 DOWN (4)

5. The product of 16 ACROSS and the difference between 1 DOWN and 20 DOWN (2)

6. Three less than 4 DOWN (3)

7. The number of digits in $25^{55} \times 1024^{11}$ (3)

9. A power of 2 (3)

12. A cube (3)

14. A prime number that is the sum of the first few consecutive prime numbers (2)

16. A factor of 10 DOWN that is a multiple of a square greater than 1 (2)

17. A multiple of 18 DOWN (3)

19. The square of a prime number, with digits in descending order (3)

21. A Fibonacci number where all adjacent digits differ by one (3)

23. 23 DOWN increased by 1130% (3)

24. A number with an odd number of factors (2)

25. The product of the first five prime numbers (4)

26. $4x + 14$ where $x = \dfrac{10\text{ DOWN}}{22} - 15\text{ DOWN}$ (3)

CROSSNUMBER

TEAM MATHS CHALLENGE
2019
REGIONAL FINAL

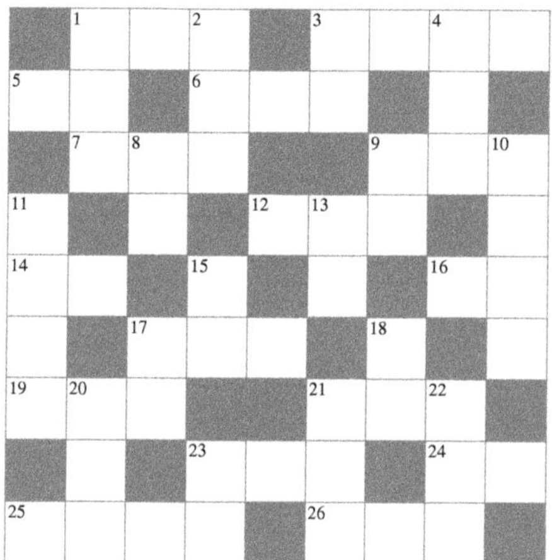

DOWN

1. A multiple of 8 DOWN (3)
2. The sum of two squares (3)
3. The sum of the first five prime numbers (2)
4. Three more than 6 ACROSS (3)
8. A factor of 1 DOWN (2)
9. A Fibonacci number divisible by 11 (2)
10. The mean of 25 ACROSS and 11 DOWN (4)
11. This number is reversed if you multiply it by 4 (4)
13. A cube (2)
15. The remainder when 21 ACROSS is divided by 9 DOWN (2)
17. An odd number that is a fourth power (2)
18. The product of two triangular numbers that is not itself a triangular number (2)
20. The number of degrees in a rhombus plus 23 ACROSS (3)
21. The difference between 21 ACROSS and 9 DOWN (3)
22. x where $3\text{ ACROSS} - \dfrac{25\text{ ACROSS}}{30} = 3x - 2$ (3)
23. The number of digits in $2 \times 125^3 \times 4^4$ (2)

Crossnumber

	[1] 4	9	[2] 6		[3] 2	2	[4] 7	7
[5] 8	8		[6] 6	9	8		0	
	[7] 1	[8] 1	1			[9] 5	1	[10] 2
[11] 2		3		[12] 1	[13] 2	5		2
[14] 1	7		[15] 5		7		[16] 4	4
7		[17] 8	2	8		[18] 1		4
[19] 8	[20] 4	1			[21] 9	8	[22] 7	
	8		[23] 1	2	3		[24] 3	6
[25] 2	3	1	0		[26] 2	1	4	

Marking Instructions—a reminder

- Pairs should write their own answers in the Answer Grid; teachers should not do this on their behalf.
- Pairs may only communicate through the teacher, and only to request that the other pair work on a particular clue.
- When a pair enters an answer in the Answer Grid, the teacher checks each digit of the answer:
  - if it is correct, place a tick in the dotted circle and award one mark
  - if it is wrong, cross it out, write in the correct digit, and place a cross in the dotted circle
  - show the correct answer to both pairs so that they are up-to-date.
- A pair may enter just one digit if they wish, rather than a complete answer.
- A pair may sacrifice a square, by guessing, if they wish.

Shuttle

United Kingdom Mathematics Trust

TEAM MATHS CHALLENGE 2019

REGIONAL FINAL

SHUTTLE

A1

Pass on the value of $24 - 16 \div 2^3$.

United Kingdom Mathematics Trust

TEAM MATHS CHALLENGE 2019

REGIONAL FINAL

SHUTTLE

A2

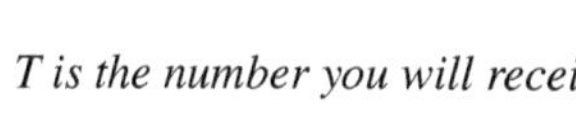

T is the number you will receive.

$\frac{T}{15} - \frac{7}{20} = \frac{a}{b}$, a fraction in its lowest terms.

Pass on the value of $a - b$.

United Kingdom Mathematics Trust

TEAM MATHS CHALLENGE 2019

REGIONAL FINAL

SHUTTLE

T is the number you will receive. **A3**

Nala, Lana and Alan share $27T$ sweets in the ratio $4:3:2$.

Nala receives K sweets.

Pass on the value of K.

United Kingdom Mathematics Trust

TEAM MATHS CHALLENGE 2019

REGIONAL FINAL

SHUTTLE

T is the number you will receive. **A4**

The first three terms in a sequence are

$$2019, \quad 2019 - T, \quad 2019 - 2T.$$

Each term is T less than the previous term.

Write down the value of the 25th term.

United Kingdom Mathematics Trust

TEAM MATHS CHALLENGE 2019

REGIONAL FINAL

SHUTTLE

B1

Pass on the value of

$$\frac{10 \times 9 \times 8}{6 \times 5 \times 4 \div (3 \times 2 \times 1)}.$$

United Kingdom Mathematics Trust

TEAM MATHS CHALLENGE 2019

REGIONAL FINAL

SHUTTLE

B2

T is the number you will receive.

$$\frac{201}{9} - 20\frac{1}{9} = \frac{K}{T}$$

Pass on the value of K.

United Kingdom Mathematics Trust

Team Maths Challenge 2019

Regional Final

Shuttle

B3

T is the number you will receive.

Ajesh and Beejesh race each other along a straight track.

Ajesh runs T metres in 10 seconds.
Beejesh runs T metres in 8 seconds.

They start together, and after K seconds, they are 60 metres apart.

Pass on the value of K.

United Kingdom Mathematics Trust

Team Maths Challenge 2019

Regional Final

Shuttle

B4

T is the number you will receive.

A sequence starts

$$2019, \quad 2018, \quad 1, \quad \ldots .$$

In this sequence, each successive term is the positive difference between the previous two terms.

Write down the value of the $(T + 1)$th term.

United Kingdom Mathematics Trust

TEAM MATHS CHALLENGE 2019

REGIONAL FINAL

SHUTTLE

C1

$144 = 2^a \times 3^b$ and $324 = 2^c \times 3^d$.

Pass on the value of $a + b + c + d - 2$.

United Kingdom Mathematics Trust

TEAM MATHS CHALLENGE 2019

REGIONAL FINAL

SHUTTLE

C2

T is the number you will receive.

Let

$$X = T\left(\frac{1}{2} + \frac{2}{3} \times \frac{3}{4} \div \frac{5}{6}\right).$$

Pass on the value of $X + 1$.

United Kingdom Mathematics Trust

Team Maths Challenge 2019

Regional Final

Shuttle

C3

T is the number you will receive.

Two regular polygons, one inside the other, share a common side. The inside polygon has $\frac{T}{2}$ sides. The outside polygon has T sides.

This is indicated in the diagram below.

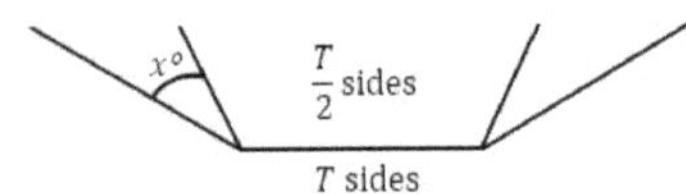

Pass on the value of x.

United Kingdom Mathematics Trust

Team Maths Challenge 2019

Regional Final

Shuttle

C4

T is the number you will receive.

Abdul says to Paula:

"If you give me T pencils, I will have twice as many pencils as you."

Paula says to Abdul:

"If you give me T pencils, I will have three times as many pencils as you."

Write down the total number of pencils that Abdul and Paula have.

United Kingdom
Mathematics Trust

Team Maths
Challenge
2019

Regional Final

Shuttle

D1

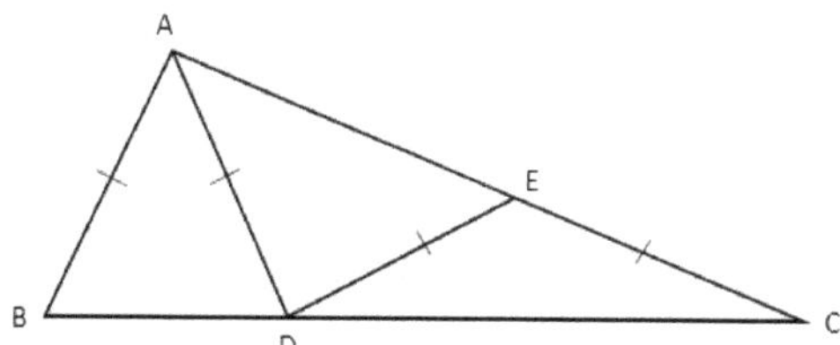

In triangle ABC, D is a point on BC such that $AB = AD$ and E is a point on AC such that $EC = ED = AD$.

Angle ABC is $60°$. Angle CAB is $K°$.

Pass on the value of K.

United Kingdom
Mathematics Trust

Team Maths
Challenge
2019

Regional Final

Shuttle

D2

T is the number you will receive.

Let $\sqrt{T}(x-3) - 2\left(x - \sqrt{T}\right) = 4\sqrt{T} - (10 - 3x)$.

Pass on the value of x.

Team Maths Challenge 2019

Regional Final

Shuttle

T is the number you will receive.

D3

Point A with coordinates $(3, 2)$ is reflected in the line $y = 4$ to obtain point B.

Point B is then reflected in the line $y = x$ to obtain point C.

Finally, point C is reflected in the line $x = T$ to obtain point D.

The coordinates of point D are (a, b).

Pass on the value of $a - 2b$.

Team Maths Challenge 2019

Regional Final

Shuttle

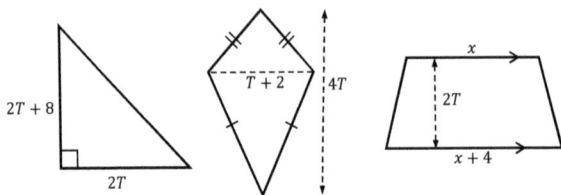

T is the number you will receive.

D4

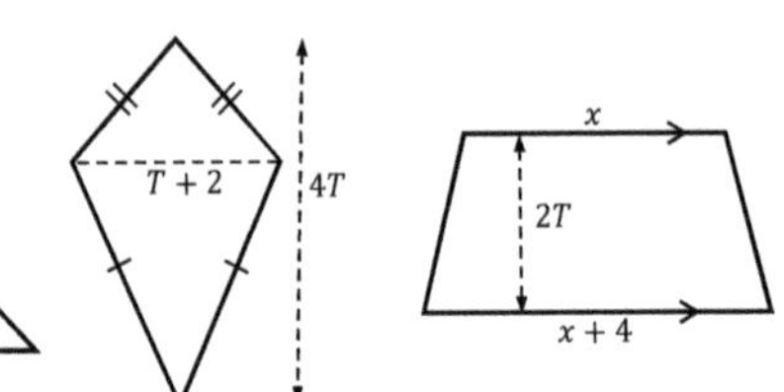

The sum of the area of the triangle and the area of the kite is equal to the area of the trapezium.

Write down the value of x.

SHUTTLE

ANSWERS

A1	B1	C1	D1
22	36	10	100
A2	**B2**	**C2**	**D2**
7	80	12	8
A3	**B3**	**C3**	**D3**
84	30	30	4
A4	**B4**	**C4**	**D4**
3	1999	144	12

On the RESPONSE SHEET: Circle the mark awarded for each question and cross out the others. At the end of the round, either circle the bonus mark or cross it out.

Relay

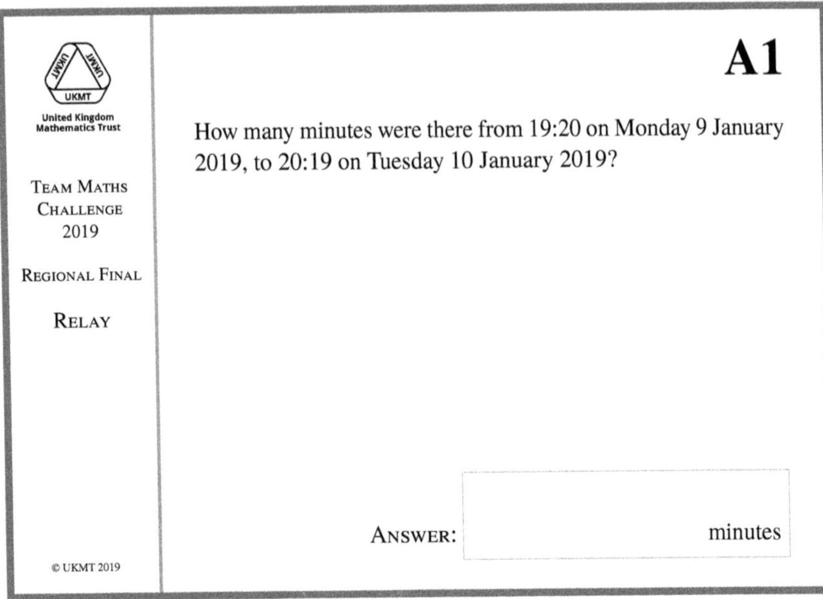

How many minutes were there from 19:20 on Monday 9 January 2019, to 20:19 on Tuesday 10 January 2019?

Answer: minutes

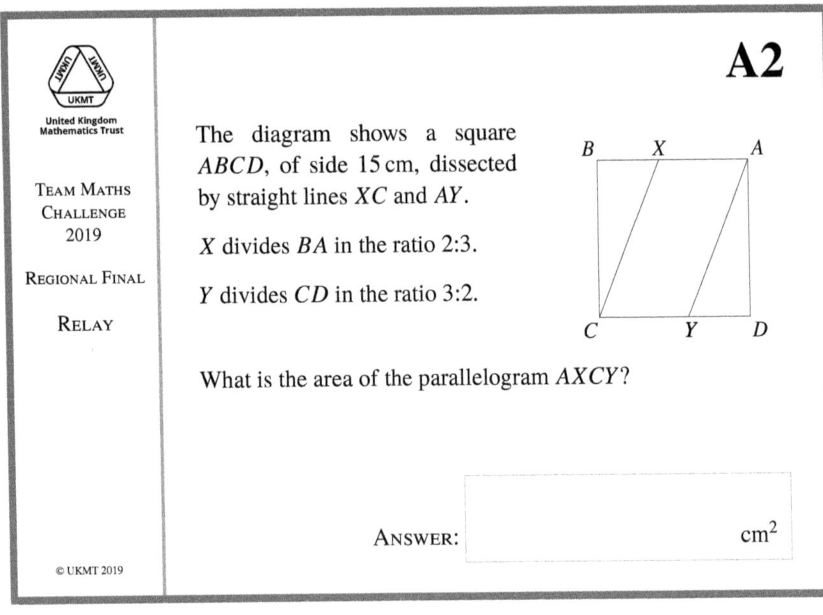

The diagram shows a square $ABCD$, of side 15 cm, dissected by straight lines XC and AY.

X divides BA in the ratio 2:3.

Y divides CD in the ratio 3:2.

What is the area of the parallelogram $AXCY$?

Answer: cm^2

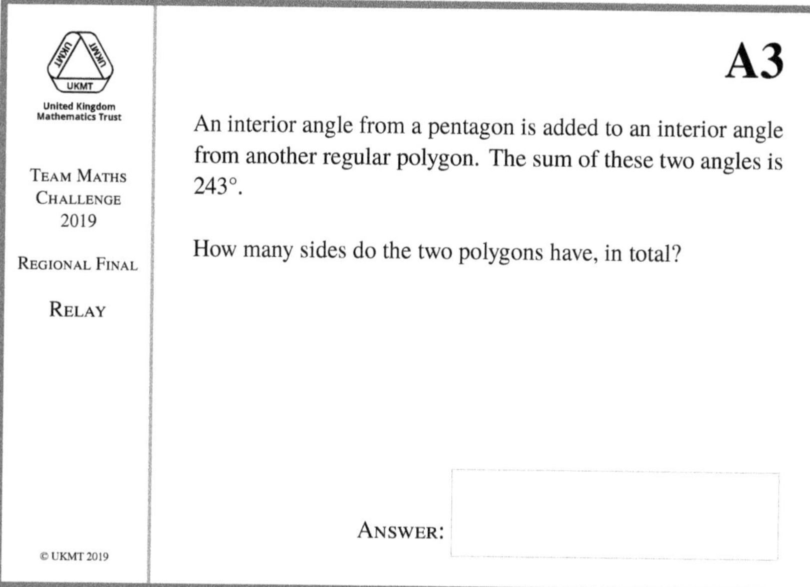
UKMT
United Kingdom Mathematics Trust
TEAM MATHS CHALLENGE 2019
REGIONAL FINAL
RELAY
© UKMT 2019
A3
An interior angle from a pentagon is added to an interior angle from another regular polygon. The sum of these two angles is 243°.
How many sides do the two polygons have, in total?
ANSWER:

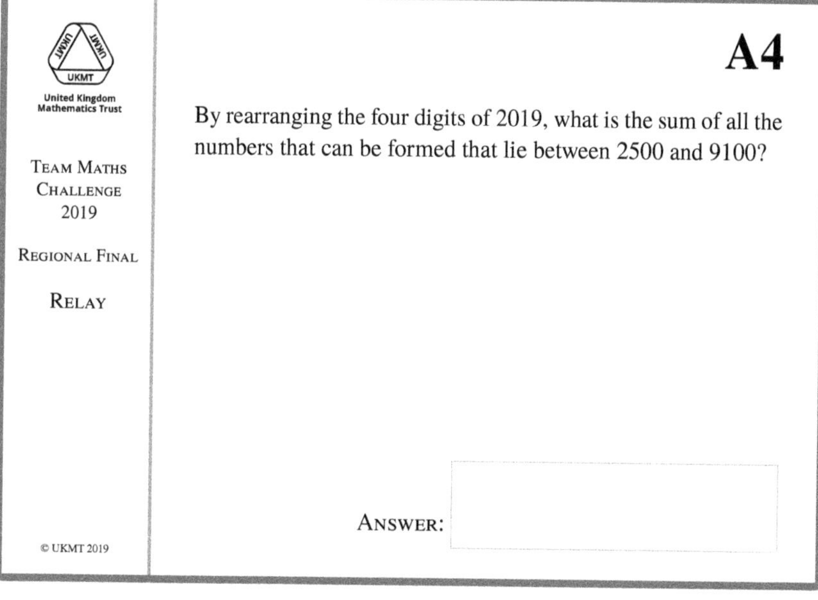
UKMT
United Kingdom Mathematics Trust
TEAM MATHS CHALLENGE 2019
REGIONAL FINAL
RELAY
© UKMT 2019
A4
By rearranging the four digits of 2019, what is the sum of all the numbers that can be formed that lie between 2500 and 9100?
ANSWER:

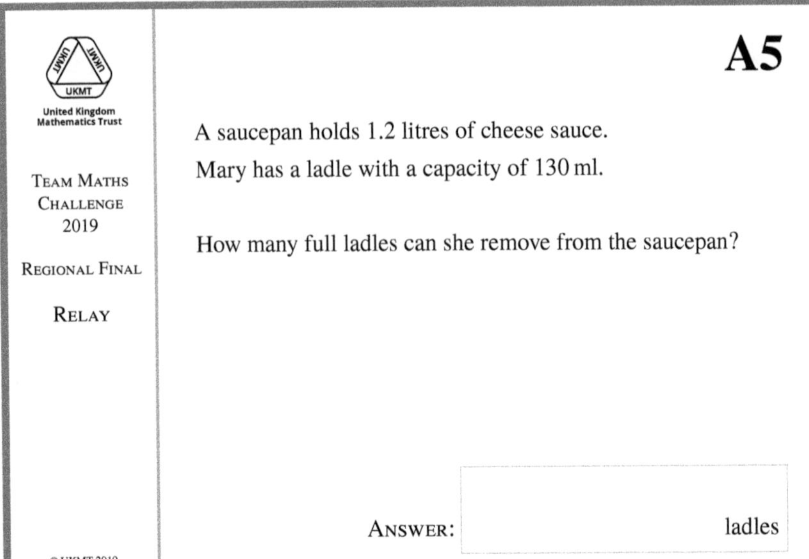

UKMT
United Kingdom Mathematics Trust

Team Maths Challenge 2019

Regional Final

Relay

A5

A saucepan holds 1.2 litres of cheese sauce.

Mary has a ladle with a capacity of 130 ml.

How many full ladles can she remove from the saucepan?

Answer: ladles

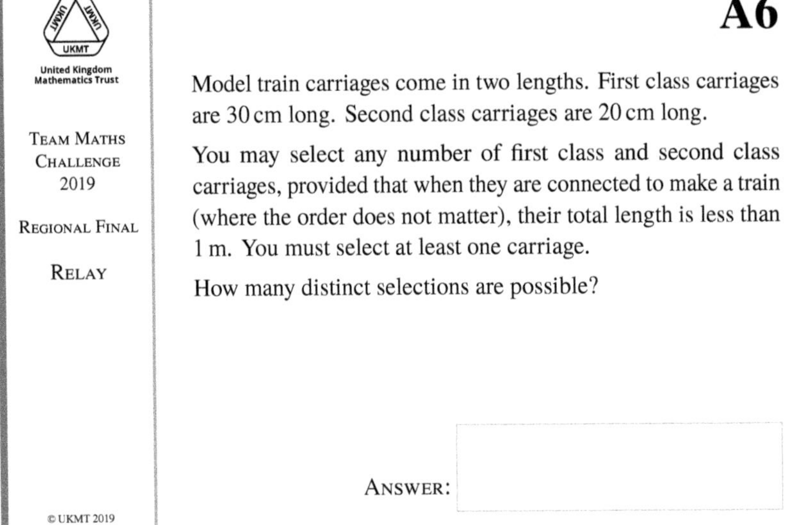

UKMT
United Kingdom Mathematics Trust

Team Maths Challenge 2019

Regional Final

Relay

A6

Model train carriages come in two lengths. First class carriages are 30 cm long. Second class carriages are 20 cm long.

You may select any number of first class and second class carriages, provided that when they are connected to make a train (where the order does not matter), their total length is less than 1 m. You must select at least one carriage.

How many distinct selections are possible?

Answer:

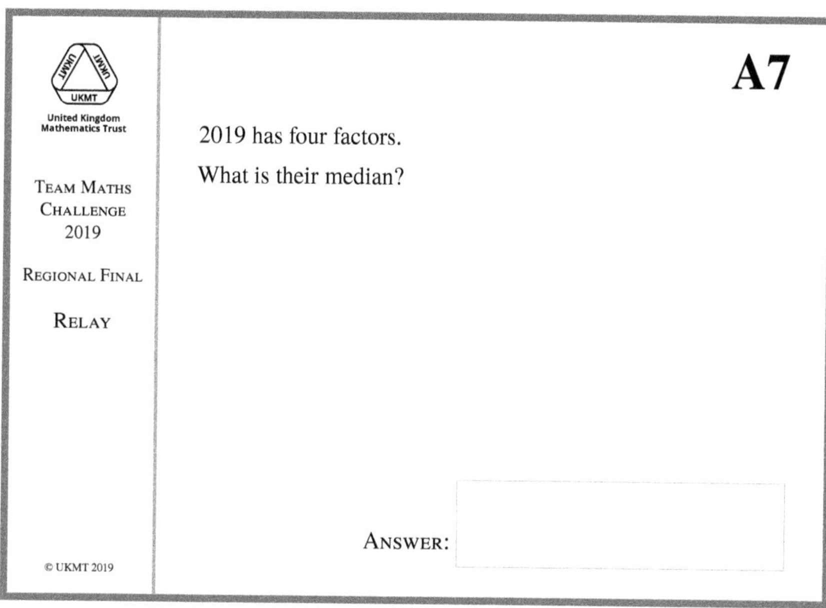
UKMT
United Kingdom Mathematics Trust
TEAM MATHS CHALLENGE 2019
REGIONAL FINAL
RELAY
A7
2019 has four factors.
What is their median?
ANSWER:
© UKMT 2019

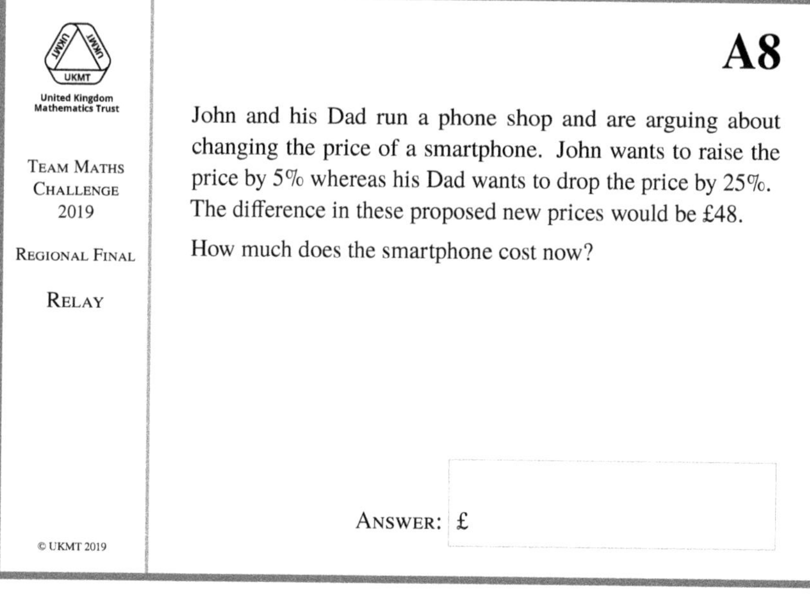
UKMT
United Kingdom Mathematics Trust
TEAM MATHS CHALLENGE 2019
REGIONAL FINAL
RELAY
A8
John and his Dad run a phone shop and are arguing about changing the price of a smartphone. John wants to raise the price by 5% whereas his Dad wants to drop the price by 25%. The difference in these proposed new prices would be £48.
How much does the smartphone cost now?
ANSWER: £
© UKMT 2019

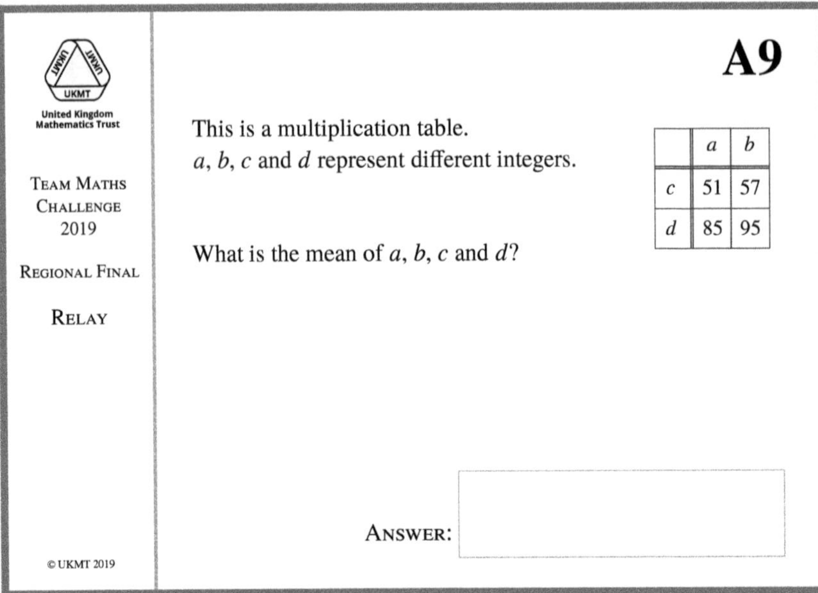

UKMT

United Kingdom Mathematics Trust

TEAM MATHS CHALLENGE 2019

REGIONAL FINAL

RELAY

A9

This is a multiplication table.
a, b, c and d represent different integers.

	a	b
c	51	57
d	85	95

What is the mean of a, b, c and d?

ANSWER:

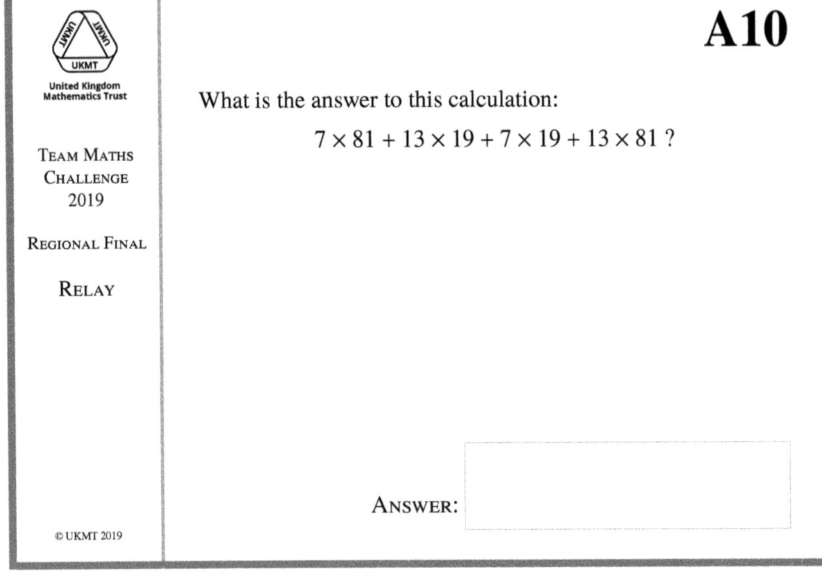

UKMT

United Kingdom Mathematics Trust

TEAM MATHS CHALLENGE 2019

REGIONAL FINAL

RELAY

A10

What is the answer to this calculation:

$$7 \times 81 + 13 \times 19 + 7 \times 19 + 13 \times 81\ ?$$

ANSWER:

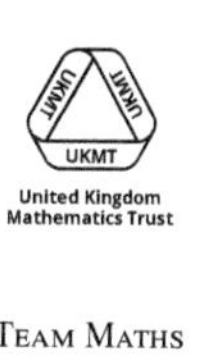

United Kingdom
Mathematics Trust

TEAM MATHS CHALLENGE 2019

REGIONAL FINAL

RELAY

A11

A trapezium has a height of 5 cm.
One of the parallel sides has a length of 12 cm.
The length of one of the parallel sides is $\frac{2}{3}$ of the length of the other parallel side.

What is the difference in the areas of the two possible trapeziums?

ANSWER: $\quad$ cm^2

United Kingdom
Mathematics Trust

TEAM MATHS CHALLENGE 2019

REGIONAL FINAL

RELAY

A12

The value of the expression $(p + q)^5$, when $p = 2$ and $q = 4$, is 7776.

What is the value of the expression $(p + q)^5$, when $p = 9$ and $q = 3$?

ANSWER:

United Kingdom Mathematics Trust

Team Maths Challenge 2019

Regional Final

Relay

A13

The diagram shows two gear wheels, A and B, meshing and rotating.

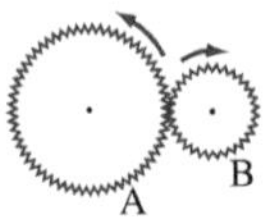

A completes 5 full rotations in $2\frac{1}{2}$ minutes

B completes 3 full rotations in 20 seconds.

What is the ratio of the number of teeth on A to the number of teeth on B. Give your answer in the form $a : b$ where a and b are positive integers with no common factor other than 1.

Answer:

United Kingdom Mathematics Trust

Team Maths Challenge 2019

Regional Final

Relay

A14

An exam paper has three sections, A, B and C.

Each section has 10 questions.

In section A, questions are worth 10 marks each.
In section B, questions are worth 20 marks each.
In section C, questions are worth 25 marks each.

You may choose any questions provided that:

- The total number of questions does not exceed 10,
- At least four questions are chosen from A,
- There are more questions chosen from B than from C.

What is the maximum possible total score on this exam paper?

Answer: marks

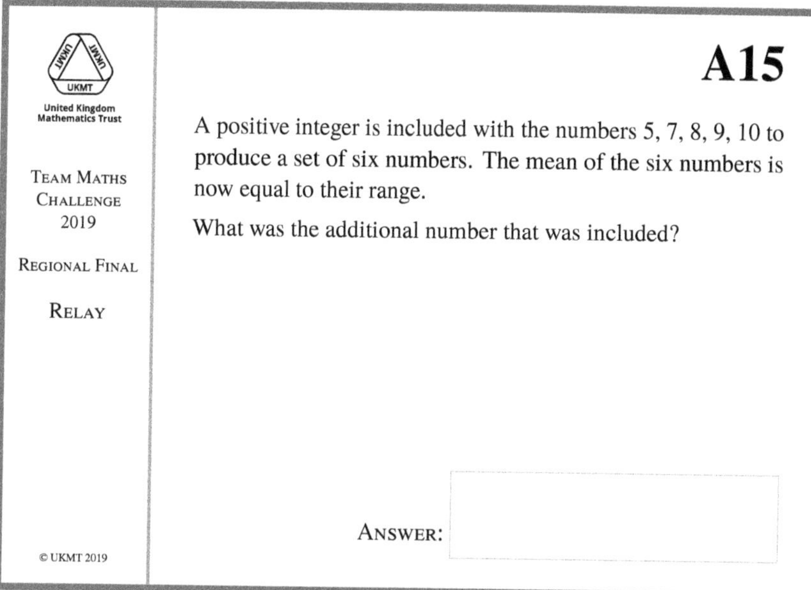

UKMT
United Kingdom Mathematics Trust

TEAM MATHS CHALLENGE 2019

REGIONAL FINAL

RELAY

A15

A positive integer is included with the numbers 5, 7, 8, 9, 10 to produce a set of six numbers. The mean of the six numbers is now equal to their range.

What was the additional number that was included?

ANSWER:

UKMT
United Kingdom Mathematics Trust

TEAM MATHS CHALLENGE 2019

REGIONAL FINAL

RELAY

B1

By rearranging the four digits of 2019, what is the sum of all the even numbers that can be formed that are greater than 9000?

ANSWER:

United Kingdom Mathematics Trust

TEAM MATHS CHALLENGE 2019

REGIONAL FINAL

RELAY

B2

John is in a race that takes him 30 minutes to complete.
After he has started, his smartwatch reports his heart rate to him every 2 min 20 sec.

How many times does his smartwatch report his heart rate, during the race?

ANSWER:

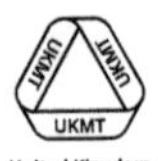

United Kingdom Mathematics Trust

TEAM MATHS CHALLENGE 2019

REGIONAL FINAL

RELAY

B3

The rectangular region R, is bounded by the lines $x = 3$, $x = 8$, $y = 4$, and $y = 6$.

How many points, whose coordinates are both integers, lie on the boundary of R?

ANSWER:

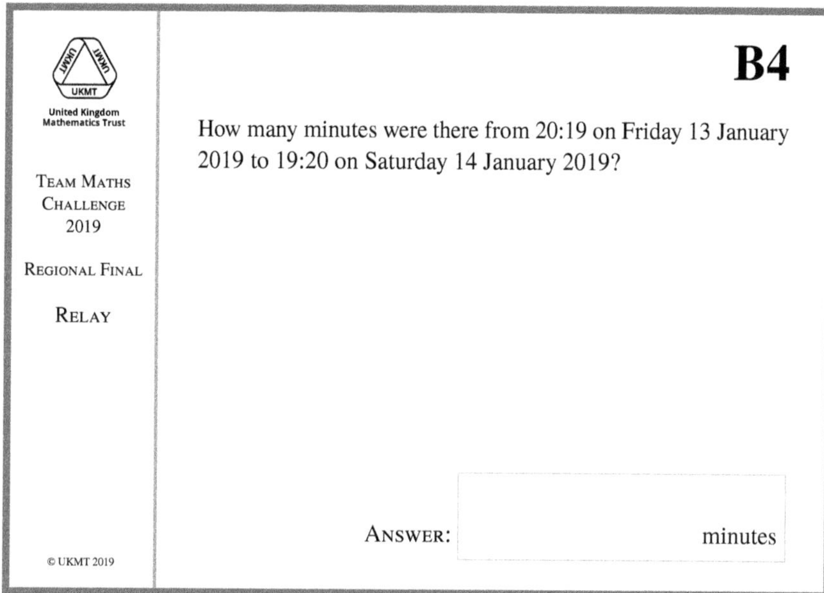
UKMT
United Kingdom Mathematics Trust
TEAM MATHS CHALLENGE 2019
REGIONAL FINAL
RELAY
B4
How many minutes were there from 20:19 on Friday 13 January 2019 to 19:20 on Saturday 14 January 2019?
ANSWER:
minutes
© UKMT 2019

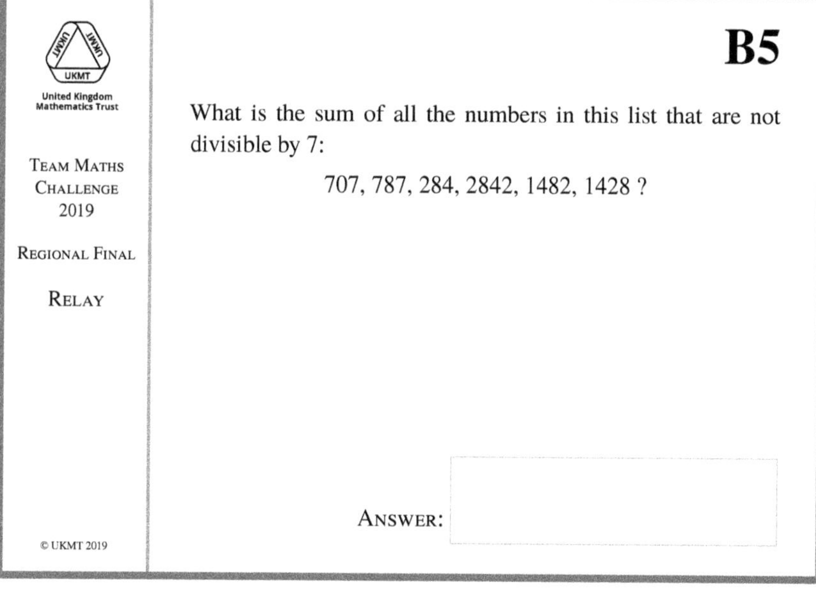
UKMT
United Kingdom Mathematics Trust
TEAM MATHS CHALLENGE 2019
REGIONAL FINAL
RELAY
B5
What is the sum of all the numbers in this list that are not divisible by 7:
707, 787, 284, 2842, 1482, 1428 ?
ANSWER:
© UKMT 2019

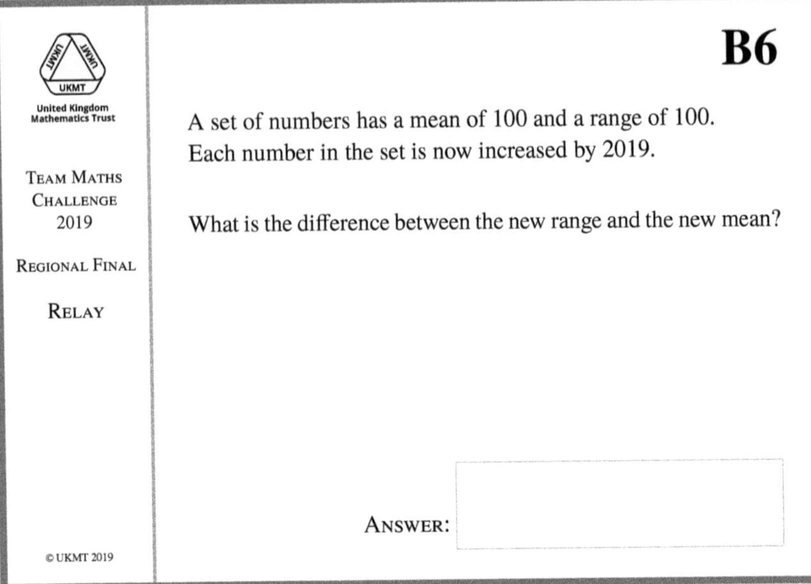

UKMT

United Kingdom Mathematics Trust

Team Maths Challenge 2019

Regional Final

Relay

B6

A set of numbers has a mean of 100 and a range of 100.
Each number in the set is now increased by 2019.

What is the difference between the new range and the new mean?

Answer:

© UKMT 2019

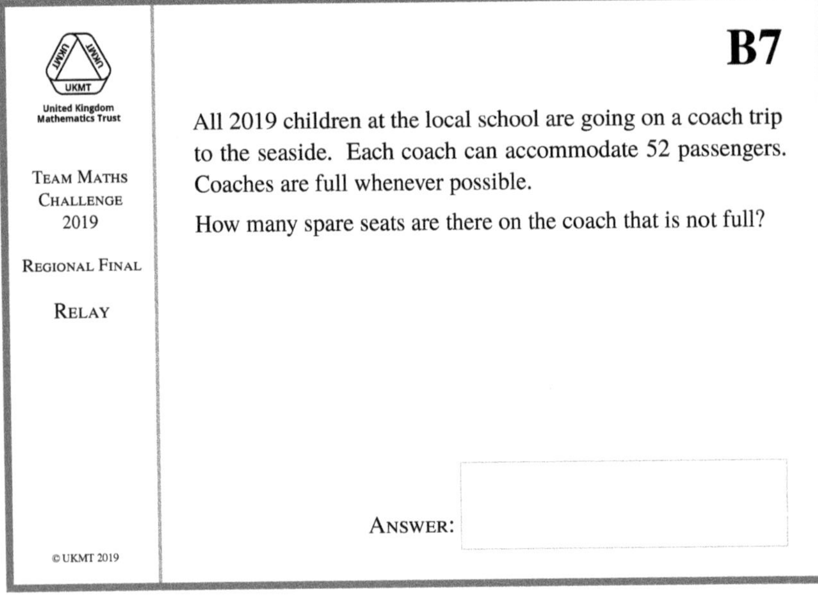

UKMT

United Kingdom Mathematics Trust

Team Maths Challenge 2019

Regional Final

Relay

B7

All 2019 children at the local school are going on a coach trip to the seaside. Each coach can accommodate 52 passengers. Coaches are full whenever possible.

How many spare seats are there on the coach that is not full?

Answer:

© UKMT 2019

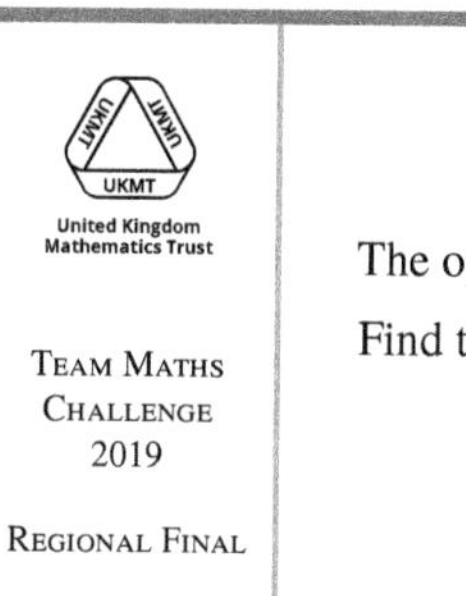

TEAM MATHS CHALLENGE 2019

REGIONAL FINAL

RELAY

B8

The operator '$\diamond$' is defined as follows: $a \diamond b = a + 2ab$

Find the value of x that makes this equation true:

$$(3 \diamond x) + (x \diamond 3) = 94$$

ANSWER:

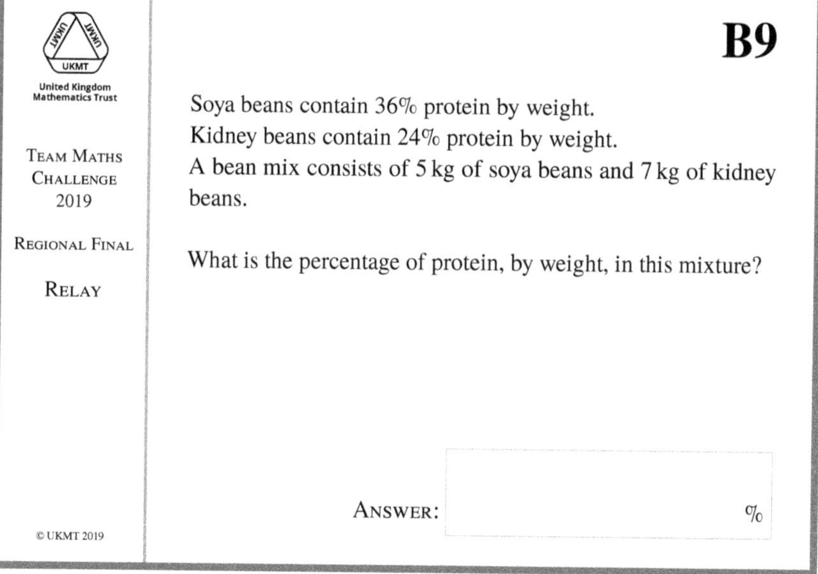

United Kingdom Mathematics Trust

TEAM MATHS CHALLENGE 2019

REGIONAL FINAL

RELAY

B9

Soya beans contain 36% protein by weight.
Kidney beans contain 24% protein by weight.
A bean mix consists of 5 kg of soya beans and 7 kg of kidney beans.

What is the percentage of protein, by weight, in this mixture?

ANSWER: %

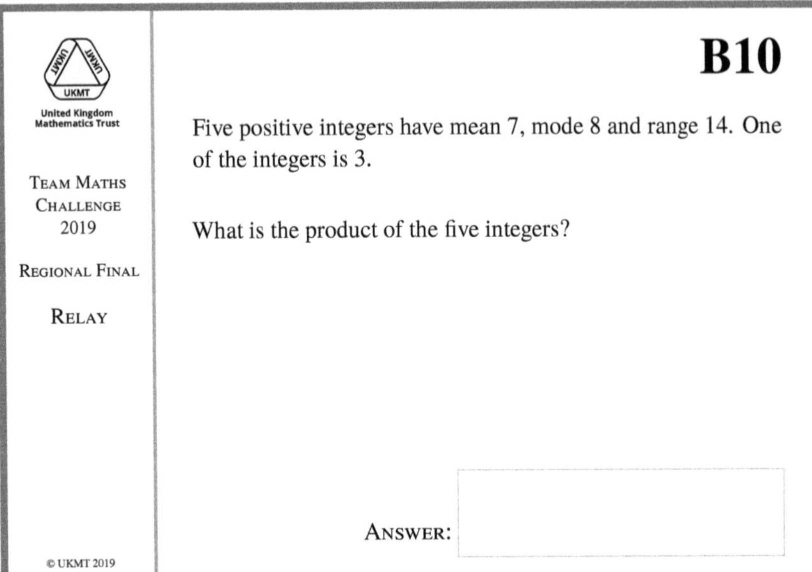
UKMT
United Kingdom Mathematics Trust
Team Maths Challenge 2019
Regional Final
Relay
B10
Five positive integers have mean 7, mode 8 and range 14. One of the integers is 3.
What is the product of the five integers?
Answer:
© UKMT 2019

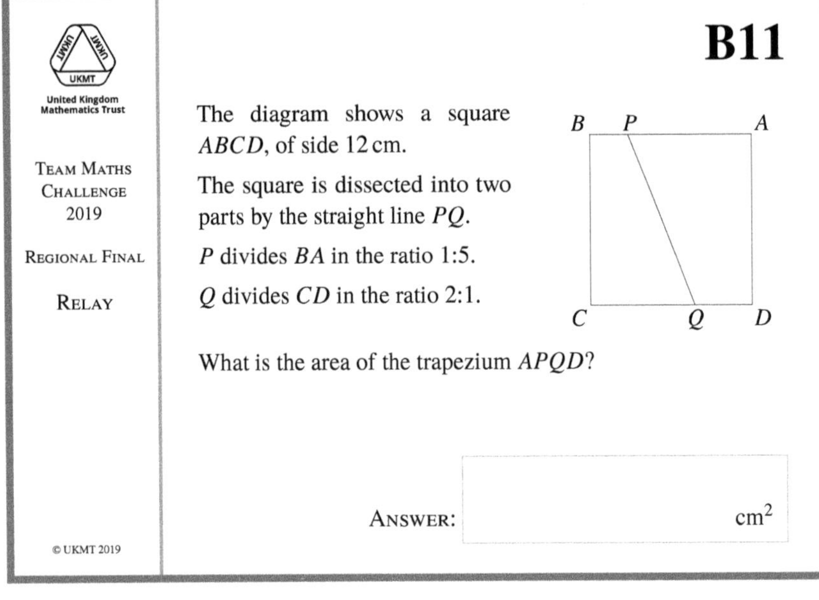
UKMT
United Kingdom Mathematics Trust
Team Maths Challenge 2019
Regional Final
Relay
B11
The diagram shows a square ABCD, of side 12 cm.
The square is dissected into two parts by the straight line PQ.
P divides BA in the ratio 1:5.
Q divides CD in the ratio 2:1.
B
P
A
C
Q
D
What is the area of the trapezium APQD?
Answer:
cm²
© UKMT 2019

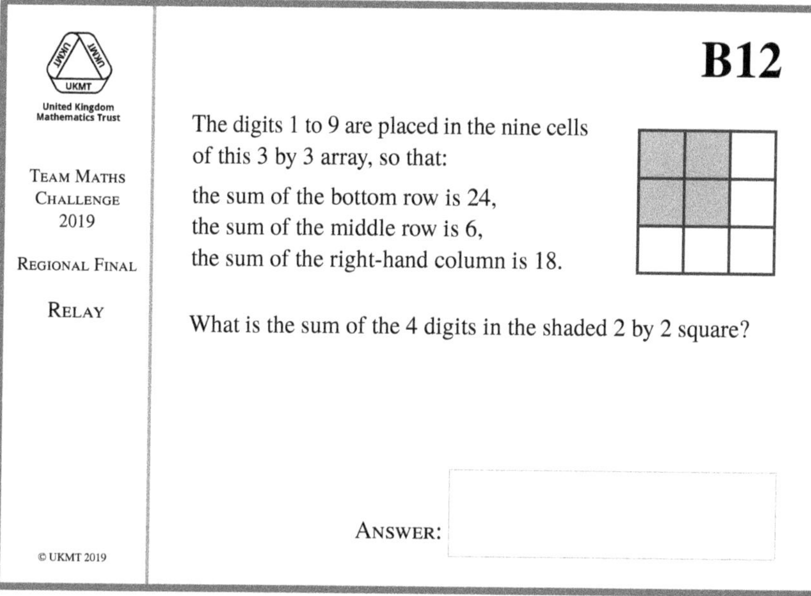
UKMT
United Kingdom Mathematics Trust
Team Maths Challenge 2019
Regional Final
Relay
B12
The digits 1 to 9 are placed in the nine cells of this 3 by 3 array, so that:
the sum of the bottom row is 24,
the sum of the middle row is 6,
the sum of the right-hand column is 18.
What is the sum of the 4 digits in the shaded 2 by 2 square?
Answer:
© UKMT 2019

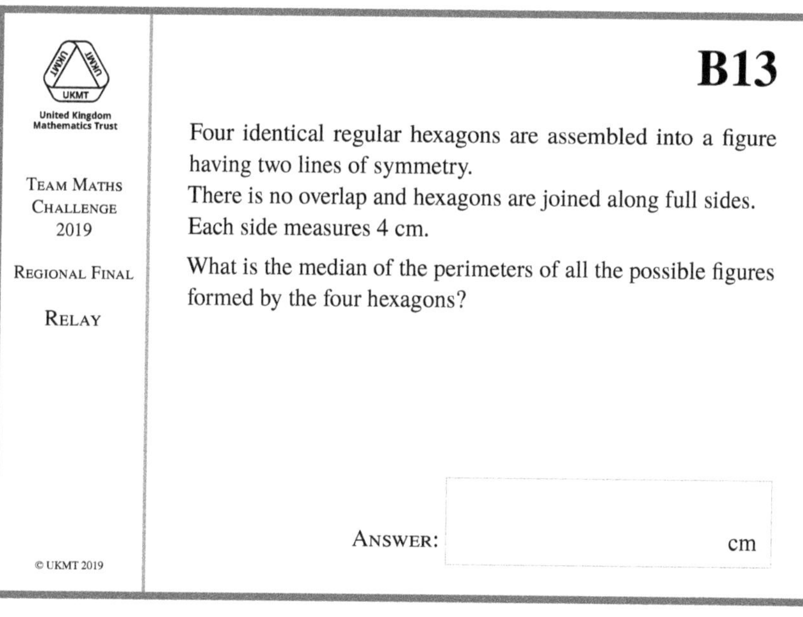
UKMT
United Kingdom Mathematics Trust
Team Maths Challenge 2019
Regional Final
Relay
B13
Four identical regular hexagons are assembled into a figure having two lines of symmetry.
There is no overlap and hexagons are joined along full sides.
Each side measures 4 cm.
What is the median of the perimeters of all the possible figures formed by the four hexagons?
Answer: cm
© UKMT 2019

United Kingdom Mathematics Trust

TEAM MATHS CHALLENGE 2019

REGIONAL FINAL

RELAY

B14

A set of traffic lights follows this cycle:

Green – Amber – Red – Red & Amber (together)

This cycle repeats indefinitely. Each cycle takes 34 seconds.

The Green light is lit for half this time. The Red light is lit for 9 seconds more than the Amber light. The Red and Amber lights are lit together for 15 seconds less than the time that the Green light is lit.

How long is the Red light lit during each cycle?

ANSWER: ______ seconds

United Kingdom Mathematics Trust

TEAM MATHS CHALLENGE 2019

REGIONAL FINAL

RELAY

B15

Sanjeev has 9 black beads and 5 white beads.
There are two trays, one square and one circular.
Sanjeev's teacher has challenged him to put the beads in the trays satisfying all of these conditions:

- All the beads must be used,
- Both trays must contain both black and white beads,
- There must be more black beads than white beads in both trays.

In how many different ways can Sanjeev carry out this task?

ANSWER: ______

RELAY

SCORESHEET

TEAM NUMBER

SCHOOL NAME

A1 1499 minutes (0)(2)	**A6** 11 (0)(2)	**A11** 25 cm^2 (0)(2)
B1 36444 (0)(2)	**B6** 2019 (0)(2)	**B11** 84 cm^2 (0)(2)
A2 135 cm^2 (0)(2)	**A7** 338 (0)(2)	**A12** 248832 (0)(2)
B2 12 (0)(2)	**B7** 9 (0)(2)	**B12** 12 (0)(2)
A3 13 (0)(2)	**A8** 160 £ (0)(2)	**A13** 9:2 (0)(2)
B3 14 (0)(2)	**B8** 7 (0)(2)	**B13** 64 cm (0)(2)
A4 23844 (0)(2)	**A9** 11 (0)(2)	**A14** 170 marks (0)(2)
B4 1381 minutes (0)(2)	**B9** 29 % (0)(2)	**B14** 14 seconds (0)(2)
A5 9 ladles (0)(2)	**A10** 2000 (0)(2)	**A15** 3 (0)(2)
B5 2553 (0)(2)	**B10** 2880 (0)(2)	**B15** 12 (0)(2)

Correct answers score 2 points: circle 2 or 0 for each question and cross out the other number.

At the end of the round, draw a line under the last question attempted.

FINAL SCORE /60

3.3 Senior Team Maths Challenge

The Senior Team Maths Challenge has the same aims as the Team Maths Challenge as noted on page 163 and is run jointly with the Further Mathematics Support Programme.
A team in the Senior Team Maths Challenge consists of four students from Years 11–13 (England and Wales), S5–S6 (Scotland) or Years 12–14 (Northern Ireland) with no more than two students from the uppermost year group.
Regional Finals take place in dozens of venues across the UK during November–December. Each team travels with an accompanying teacher and competes against teams from other schools in the area in a series of four rounds. The 2019–20 competition had 1174 teams competing in 68 Regional Finals.

Group Round The team has 40 minutes to answer 10 questions. It is for the team to decide on a strategy to tackle the problems.

Crossnumber The time allowed for this round is 40 minutes. Teams divide into pairs and each pair works independently on either the ACROSS clues or the DOWN clues. The pairs may only communicate with each other by entering digits into the grid or by asking the other pair to try to solve a particular clue.

Shuttle Teams split into pairs. One pair is given Questions 1 and 3; the other pair is given Questions 2 and 4. The first pair works on Question 1 and passes the answer to the other pair who use it to help answer Question 2. This continues with the second pair passing the answer back to the first pair, and so on, until a full set of answers is provided for marking. Bonus points are awarded to teams which present a correct set of answers before the six-minute whistle, then the other teams have another two minutes to finish. Four such shuttles are attempted in this round.

High-scoring teams from the Regional Finals are invited to compete against each other at the National Final held in London in February. All participants receive a certificate and goody bag, and the winners are presented with prizes and the Senior Team Maths Challenge trophy.

The rounds in the National Final are the same as the Regional Finals, except for the following.

Poster Competition This is an additional round where teams are given 50 minutes to design a poster on a given topic.

Relay This additional round works in a similar way to the Relay in the Teams Maths Challenge (see page 164).
The 2019–20 National Final took place at the Royal Horticultural Halls in London on Tuesday 4th February. Teams from the following 88 schools were invited.

Ackworth School
Altrincham Grammar School for Boys
Aylesbury Grammar School
Bancroft's School
Bedes Senior School
Bedford School
Bellerbys Brighton
Belper School
Birkenhead School
Bristol Grammar School
Cardiff High School
Cheltenham College
Cheltenham Ladies' College
City of London Freemens School
City of London School
Colchester Sixth Form College
Concord College
Dame Alice Owen's School
Devonport High School for Boys
Durham School
East Norfolk Multi Academy Trust (East Norfolk Sixth Form College)
EF Academy Torbay
Eton College
Exeter Maths School
Greenhead College
Haberdashers' Aske's Boys' School
Hampton School
Harris Westminster Sixth Form
Harrogate Grammar Schoool
Hereford Cathedral
High School of Dundee
High Storrs School
Hills Road Sixth Form College
Horsforth School
Hutchesons' Grammar School
INTO UEA, Norwich
Kendrick School
Kimbolton School
Kind Edward VI Grammar School
King Edward VI Camp Hill School for Girls
King Edward VI High School for Girls
King Edward VII
King's College London Maths School
Leicester Grammar School
Lime House School
Liverpool Blue Coat School
Loreto College
Merchiston Castle School
Millfield
MPW College Birmingham
Newcastle Sixth Form College
Oxford International College
Parmiter's School
Peter Symonds College
Queen Elizabeth Sixth Form College
Queen Elizabeth's Grammar School
RGS Newcastle
Ripley St Thomas Church of England Academy
Royal Grammar School (Guildford)
Runshaw College
Ruthin School
Scarborough Sixth Form College
Sevenoaks School
Sherborne School
Simon Langton Boys School
Solihull School

St Michaels
St Olave's Grammar School
St Paul's
St Paul's Girls' School
The Corsham School
The High School of Glasgow
The Manchester Grammar School
The Portsmouth Grammar School
The Tiffin Girls' School
The West Bridgford School
Thomas Telford School
Tiffin School
Tonbridge School
Truro College
Uppingham School
UWC Atlantic
Wallace High School
Westminster School
Winchester College
Withington Girls' School
Wycombe Abbey
Ysgol Friars

After a close competition, the top-placed teams were:

1. First Place: Westminster School

2. Second Place: Ruthin School

3. Third Place: Winchester College

The Poster Competition was won by Bancroft's School.

3.3.1 Regional Final materials

Group Round

GROUP ROUND

SENIOR TEAM MATHS CHALLENGE
2019–20
REGIONAL FINAL

Instructions

- Your team will have 40 minutes to answer 10 questions. Each team will have the same questions.
- Each question is worth 6 points. However, some questions are easier than others.
- You will have to decide your team's strategy for this group competition. Do you split up so that individuals work on a few questions each or do you work in pairs on a greater number of questions? Working all together on all the questions may well take too long.
- There is only one RESPONSE SHEET per team. Five minutes before the end of the time you will be told to finalise your answers and write them on the RESPONSE SHEET. Only this RESPONSE SHEET will be marked.

GROUP ROUND

SENIOR TEAM MATHS CHALLENGE
2019–20
REGIONAL FINAL

QUESTION 1

Find the value of

$$\left(\left((2+0)^{-1}+1\right)^{-1}+9\right)^{-1}.$$

Give your answer in the form $\frac{p}{q}$, where p and q are prime numbers.

[6 marks]

GROUP ROUND

SENIOR TEAM MATHS CHALLENGE
2019–20
REGIONAL FINAL

QUESTION 2

Annabel, Brian, Clodagh and Davinder spent a week in the summer picking grapes.

They all worked out the total mass of grapes that they had each collected. It was found that Annabel had picked 20% more grapes (by mass) than Brian and the mass of grapes brought in by Brian was 20% less than the mass collected by Clodagh. The masses of Clodagh and Davinder's hauls were in the ratio 3 : 2.

Davinder had picked 100kg of grapes.

What was the total mass, in kilograms, of fruit picked by these four people?

[6 marks]

GROUP ROUND

SENIOR TEAM MATHS CHALLENGE 2019–20 REGIONAL FINAL

QUESTION 3

2019 can be written as the product of two prime numbers and also as the the sum of two prime numbers, since $2019 = 3 \times 673$ and $2019 = 2 + 2017$.

What is the least positive integer that can be written as a product of two, not necessarily different, prime numbers but **cannot** be written as a sum of two, not necessarily different, prime numbers?

[6 marks]

Group Round

Senior Team Maths Challenge
2019–20
Regional Final

Question 4

There are 1200 pupils in a school in Venntown. 60% of the pupils study Spanish, 50% of the pupils study French but 10% of the pupils study neither language.

How many pupils study both French and Spanish?

[6 marks]

GROUP ROUND

SENIOR TEAM MATHS CHALLENGE
2019–20
REGIONAL FINAL

QUESTION 5

The sequence of Perrin numbers is defined by:

$$u_1 = 3, u_2 = 0, u_3 = 2,$$

and

$$u_{n+3} = u_{n+1} + u_n, \quad \text{for} \quad n \geq 1.$$

How many terms in the sequence are less than 100?

[6 marks]

GROUP ROUND

SENIOR TEAM MATHS CHALLENGE 2019–20 REGIONAL FINAL

QUESTION 6

The area of the rectangle $ABCD$ is equal to the area of the triangle ABE.

The lines AE and BE intersect the line CD at points F and G respectively.

The lengths of the line segments AF, BG and DF are 15, 13 and 9 respectively.

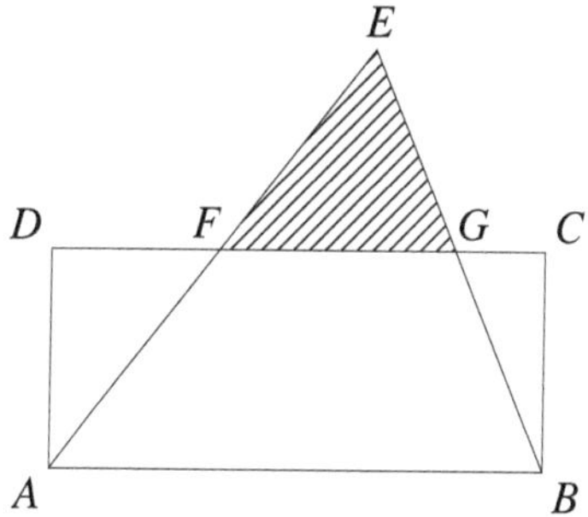

What is the area of the triangle EFG?

[6 marks]

GROUP ROUND

SENIOR TEAM MATHS CHALLENGE 2019–20 REGIONAL FINAL

QUESTION 7

In this question the letters F, O, R, T, U, N and E represent different digits.

The digits O, N, E, F, U and R satisfy

$$\begin{array}{r} O\ N\ E \\ O\ N\ E \\ O\ N\ E \\ +\ O\ N\ E \\ \hline F\ O\ U\ R \end{array}$$

N represents 5.

The digit represented by T is less than the digit represented by O.

What is the value of $F + O + U + R + T + E + E + N$?

[6 marks]

GROUP ROUND

SENIOR TEAM MATHS CHALLENGE 2019–20 REGIONAL FINAL

QUESTION 8

The diagram shows 12 small circles of radius 1 and a large circle, inside a square.

Each side of the square is a tangent to the large circle and four of the small circles.

Each small circle touches two other circles.

What is the length of each side of the square?

[6 marks]

GROUP ROUND

SENIOR TEAM MATHS CHALLENGE
2019–20
REGIONAL FINAL

QUESTION 9

A *checkmark* number is a four-digit integer '$abcd$', where a, b, c and d are digits such that $b < c < d$ and $a > b$.

2019 is an example of a *checkmark* number.

Including 2019, how many *checkmark* numbers are there?

[6 marks]

GROUP ROUND

SENIOR TEAM MATHS CHALLENGE 2019–20 REGIONAL FINAL

QUESTION 10

The diagram below shows a net of a square-based pyramid. Each edge of the pyramid has length 6 cm.

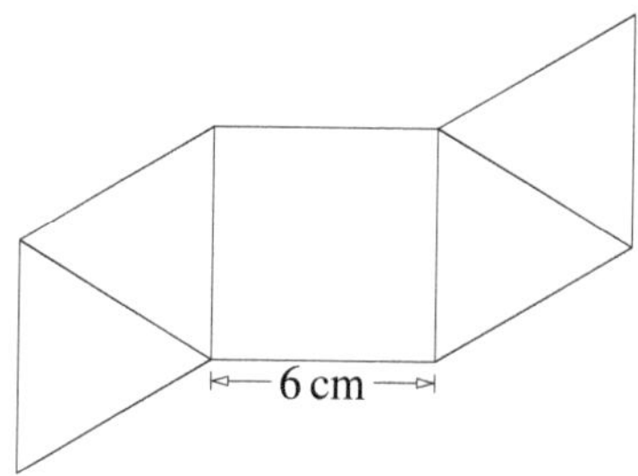

What is the volume of the pyramid, in cm^3?

Give your answer in the form $a\sqrt{p}$ where a is an integer and p is a prime number.

Note: The volume, V, of a pyramid is given by the formula

$$V = \frac{1}{3}bh$$

where b is the area of the base of the pyramid and h is the height of the pyramid.

[6 marks]

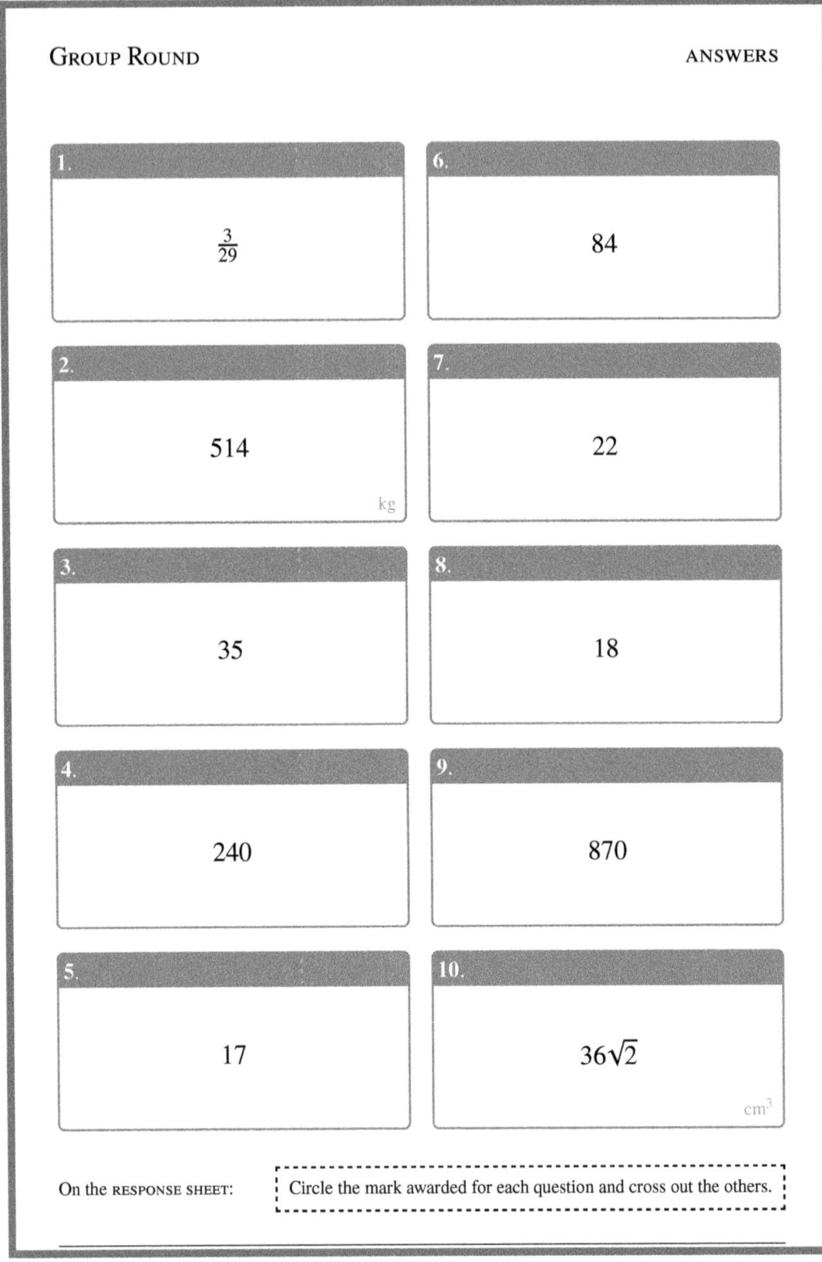

GROUP ROUND ANSWERS

1. $\frac{3}{29}$

2. 514 kg

3. 35

4. 240

5. 17

6. 84

7. 22

8. 18

9. 870

10. $36\sqrt{2}$ cm^3

On the RESPONSE SHEET: Circle the mark awarded for each question and cross out the others.

Crossnumber

CROSSNUMBER

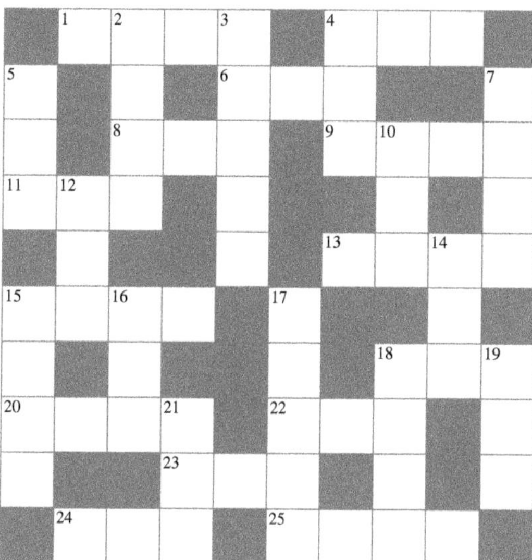

ACROSS

1. The mean of 18 DOWN, 4 DOWN and 8 ACROSS (4)
4. $(19 - \sqrt{20})(19 + \sqrt{20})$ (3)
6. A multiple of the two-digit factor of 18 ACROSS (3)
8. A Fibonacci number (3)
9. The distance between (0, 110 − 14 DOWN) and $(100\sqrt{8 \text{ ACROSS}}, 0)$ (4)
11. The number of different arrangements of the letters in the word ELEVEN (3)
13. A multiple of 8 ACROSS (4)
15. $1^4 + 6^4 + 3^4 + 4^4$ (4)
18. The number of digits in $3125^{160} \times 256^{100}$ (3)
20. A multiple of 11 (4)
22. The sum of 4 ACROSS, 24 ACROSS and 19 DOWN (3)
23. The difference between 22 ACROSS and 21 DOWN (3)
24. The product of the solutions of $m^2 - 42m + 4 \text{ ACROSS} = 0$ (3)
25. One third of the sum of the internal angles, in degrees, in a polygon with 11 ACROSS sides (4)

DOWN

2. A multiple of 12 DOWN (4)
3. The sum of 18 DOWN and twice 15 ACROSS (5)
4. $(20 - \sqrt{19})(20 + \sqrt{19})$ (3)
5. The difference between 23 ACROSS and 19 DOWN (3)
7. One sixtieth of the sum of the internal angles, in degrees, in a polygon with 10 DOWN sides (4)
10. The number of different arrangements of the letters in the word THIRTY (3)
12. A factor of 2 DOWN (3)
14. A Fibonacci number (3)
15. The area of a right-angled triangle where the two longest sides are 8 ACROSS and 145 (4)
16. The mean of 8 ACROSS and 14 DOWN (3)
17. The sum of 3 DOWN and 4 ACROSS (5)
18. $8^4 + 2^4 + 0^4 + 8^4$ (4)
19. The product of the solutions of $n^2 - 29n + 11 \text{ ACROSS} = 0$ (3)
21. The number of digits in $1024^{16} \times 625^{40}$ (3)

Crossnumber

	[1] 2	[2] 9	1	[3] 1		[4] 3	4	1	
[5] 5		3		[6] 1	7	8			[7] 1
2		[8] 1	4	4		[9] 1	[10] 3	0	0
[11] 1	[12] 2	0		7			6		7
	6			6		[13] 8	0	[14] 6	4
[15] 1	6	[16] 3	4		[17] 1			1	
2		7			1		[18] 8	0	[19] 1
[20] 2	8	7	[21] 1		[22] 8	0	2		2
4			[23] 6	4	1		0		0
	[24] 3	4	1		[25] 7	0	8	0	

Shuttle

United Kingdom Mathematics Trust

Advanced Mathematics Support Programme

Senior Team Maths Challenge 2019–20

Regional Final

Shuttle

A1

The angles of a quadrilateral are 150°, 120°, x° and y°.

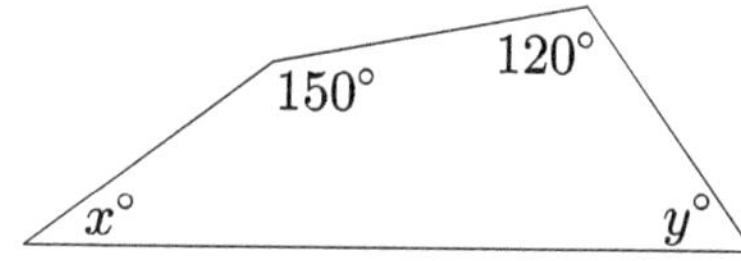

The ratio $x : y$ is $2 : 3$.

Pass on the value of x.

United Kingdom Mathematics Trust

Advanced Mathematics Support Programme

Senior Team Maths Challenge 2019–20

Regional Final

Shuttle

A3

T is the number you will receive.

David has lost some of his marbles. He started with $40T$ marbles, coloured red and blue in the ratio $5 : 4$. After losing some red marbles, the ratio is now $4 : 5$.

Pass on the number of red marbles that he lost.

United Kingdom Mathematics Trust

SENIOR TEAM MATHS CHALLENGE 2019–20

REGIONAL FINAL

SHUTTLE

A2

T is the number you will receive.

In the quadrilateral $ABCD$, $AB = AD$ and $BD = BC$.

ABE is a straight line and $\angle DAB = \angle CBE = T°$.

$\angle BDC = x°$.

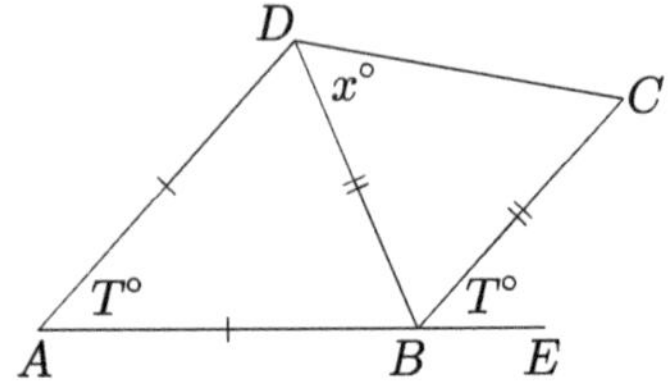

Pass on the sum of the digits of x.

United Kingdom Mathematics Trust

SENIOR TEAM MATHS CHALLENGE 2019–20

REGIONAL FINAL

SHUTTLE

A4

T is the number you will receive.

In the quadrilateral $ADCB$, the point E lies on AD so that $\angle EBA = \angle BEC = \angle DCE = 90°$.

Also $CD = 16\,\text{cm}$, $BC = 15\,\text{cm}$ and $BE = \frac{1}{8}T\,\text{cm}$.

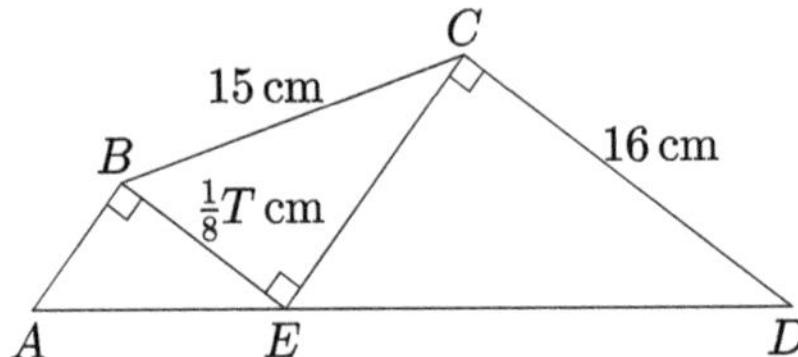

Write down the length, in cm, of AE.

Give your answer in the form $\frac{a}{b}$, where a and b are positive integers with no common factor other than 1.

United Kingdom Mathematics Trust

SENIOR TEAM MATHS CHALLENGE 2019–20

REGIONAL FINAL

SHUTTLE

B1

The positive integer 12 can be written as the sum of three consecutive positive integers: $12 = 3 + 4 + 5$.

Pass on the largest positive integer N such that 42 can be written as the sum of N consecutive positive integers.

United Kingdom Mathematics Trust

SENIOR TEAM MATHS CHALLENGE 2019–20

REGIONAL FINAL

SHUTTLE

© UKMT 2019–20

B3

T is the number you will receive.

There are $T + 5$ counters in a bag, coloured red and blue in the ratio $5 : 2$.

Equal numbers of red and blue counters are added to the bag until the ratio is $3 : 2$.

Pass on the number of red counters that are added.

United Kingdom Mathematics Trust

Advanced Mathematics Support Programme

SENIOR TEAM MATHS CHALLENGE 2019–20

REGIONAL FINAL

SHUTTLE

B2

T is the number you will receive.

The equation

$$x^2 - (7T - 2)x + 420 = 0$$

has solutions a and b with $a > b$.

Pass on the value of $a - b$.

United Kingdom Mathematics Trust

Advanced Mathematics Support Programme

SENIOR TEAM MATHS CHALLENGE 2019–20

REGIONAL FINAL

SHUTTLE

B4

T is the number you will receive.

A bag contains $2T - 2$ counters. Of these $T + 9$ are red. The probability of selecting two counters at random, of which exactly one is red, is $\frac{a}{b}$, where a and b are positive integers with no common factor other than 1.

Write down the value of $a + b$.

United Kingdom Mathematics Trust

SENIOR TEAM MATHS CHALLENGE 2019–20

REGIONAL FINAL

SHUTTLE

C1

The operation $\diamond$ is defined by

$$x \diamond y = x^2 - y.$$

Pass on the value of

$$(9 \diamond 14) - (8 \diamond 9).$$

United Kingdom Mathematics Trust

SENIOR TEAM MATHS CHALLENGE 2019–20

REGIONAL FINAL

SHUTTLE

© UKMT 2019–20

C3

T is the number you will receive.

The interior angles, in degrees, of a polygon with $\frac{2T}{105}$ sides form an arithmetic sequence with common difference $\frac{4T}{105}$.

Pass on the size, in degrees, of the smallest angle in the polygon.

Note: an *arithmetic sequence* is a sequence in which there is a common difference between consecutive terms. For example

$$2, 9, 16, 23, 30$$

is an arithmetic sequence with common difference 7.

United Kingdom Mathematics Trust

SENIOR TEAM MATHS CHALLENGE 2019–20

REGIONAL FINAL

SHUTTLE

C2

T is the number you will receive.

The volume of a cuboid is $64T$ cm^3. Its surface area is 560 cm^2.

The volume of a smaller but geometrically similar cuboid is 324 cm^3. The surface area of the smaller cuboid is S cm^2.

Pass on the value of S.

United Kingdom Mathematics Trust

SENIOR TEAM MATHS CHALLENGE 2019–20

REGIONAL FINAL

SHUTTLE

C4

T is the number you will receive.

The line with the equation

$$y = 2x + A$$

meets the circle with the equation

$$x^2 + y^2 = T$$

at just one point.

Write down the value of A^2.

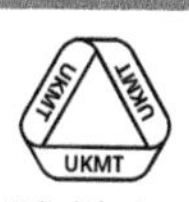

United Kingdom Mathematics Trust

Senior Team Maths Challenge 2019–20

Regional Final

Shuttle

D1

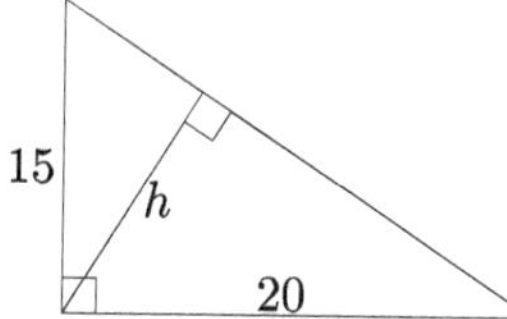

Pass on the value of h.

United Kingdom Mathematics Trust

Senior Team Maths Challenge 2019–20

Regional Final

Shuttle

D3

T is the number you will receive.

A point has coordinates $(-T - 3, 1 - 2T)$.

After reflection in the line $y = -x$, followed by a translation of 10 units to the right and 5 units up, it has coordinates (a, b).

Pass on the value of $a + b$.

United Kingdom Mathematics Trust

SENIOR TEAM MATHS CHALLENGE 2019–20

REGIONAL FINAL

SHUTTLE

D2

T is the number you will receive.

The line with the equation

$$4y = 5x + T$$

meets the parabola with the equation

$$16y = 5x^2 - 100 - T$$

at the points with coordinates (a, b) and (c, d).

Pass on the value of $a + c$.

United Kingdom Mathematics Trust

SENIOR TEAM MATHS CHALLENGE 2019–20

REGIONAL FINAL

SHUTTLE

D4

T is the number you will receive.

The integer

$$T^2 + 31T + 108$$

has n different prime factors.

Write down the value of n.

SHUTTLE ANSWERS

A1	B1	C1	D1
36	7	12	12
A2	**B2**	**C2**	**D2**
9	23	315	4
A3	**B3**	**C3**	**D3**
72	16	90	29
A4	**B4**	**C4**	**D4**
$\frac{45}{4}$	112	450	4

On the RESPONSE SHEET: Circle the mark awarded for each question and cross out the others. At the end of the round, either circle the bonus mark or cross it out.

 SENIOR TEAM MATHS CHALLENGE 2019–20 REGIONAL FINAL

3.3.2 National Final materials

Poster Competition

POSTER COMPETITION

SENIOR TEAM MATHS CHALLENGE
2020
NATIONAL FINAL

United Kingdom Mathematics Trust

Instructions

In addition to three rounds similar to those at the regional finals (Group Round, Crossnumber and Shuttle), and a Relay Round, there will be a Poster Competition at the National Final. All teams are required to take part and submit a poster.

For the Poster Competition, each team will be awarded up to 6 marks which will be added to their overall score in the main competition. The mark scheme will be as follows: 6 marks for posters on the judges' shortlist, 4 marks for posters on the judges' longlist, and 2 marks where the judges find discernible effort.

The Poster Competition also offers a chance to win a separate trophy.

After the competition some posters may be retained by the UKMT in order to be reproduced for promotional purposes.

On the day, teams will have 50 minutes to create a poster on a sheet of A1 paper (landscape), which will be provided. Sheets of A4 paper will also be available.

The subject of the poster will be *Structure and symmetry in chemistry* (see enclosed sheet). Teams must carry out research into this topic in the weeks leading up to the final.

Teams may create materials beforehand, but such prepared work must be on sheets no larger than A4 and must be assembled to create the poster on the day.

A team which arrives with a poster already assembled will be disqualified.

The materials of the poster must not extend beyond the edge of the A1 paper.

The judges will not touch the poster, so all information must be clearly visible.

Your team number (assigned to you on arrival) must be clearly visible in the bottom right-hand corner of the poster. There must be nothing else on the poster to identify the team.

Reference books may not be used at the competition, and large extracts copied directly from books or the internet will not receive much credit.

Teams must bring with them any drawing equipment they think they will need.

Glue sticks and scissors will be provided.

The content of each poster is limited only by the imagination of the team members. However, on the day each team will be presented with three questions on the subject—*the answers to these questions must be incorporated into the structure of the poster.* Teams may be asked to provide geometric or algebraic proofs, and some ingenuity may be involved.

Posters will be judged on the grade descriptors overleaf.

POSTER COMPETITION

SENIOR TEAM MATHS CHALLENGE
2020
NATIONAL FINAL

Grade descriptors

Marks will be added to each team's score in the main competition according to the following grade descriptors. In order to be awarded a particular mark, teams will generally be expected to reach the minimum standard for that mark on all three criteria, but the judges may exercise some flexibility for posters that are particularly strong on one or two out of the three criteria.

Mark	**Criterion**		
	Questions	**Mathematical content**	**Design**
6	Good attempt at some or all questions.	Mathematical content is relevant, well-chosen, clearly explained and not copied verbatim.	Poster has an eye-catching structure with potential for creating a professional final product, although execution may not be exceptional.
4	A reasonable attempt at some or all questions.	Mathematical content is included and demonstrates some understanding, rather than being reproduced entirely from other sources.	Elements of the poster are well designed and may show originality and creativity, though the constituent parts do not necessarily fit together to form a satisfactory whole.
2	Evidence that at least one of the questions has been seriously attempted.	Some mathematical content is included.	Some creativity *may* be seen, but little thought is given to overall design.
0	No discernible effort is shown.		

The separate trophy for the Poster Competition will be awarded to the poster which is awarded 6 marks towards the score in the main competition, *and* which most fully meets the above criteria.

POSTER COMPETITION QUESTIONS

TEAM MATHS CHALLENGE 2020 NATIONAL FINAL

Structure and symmetry in chemistry

Question 1

A *fullerene* is a form of carbon, each of whose molecules resembles a convex polyhedron whose faces are all pentagons, hexagons, or heptagons. Each vertex of the polyhedron is adjacent to exactly three faces. Prove that if all of the faces are pentagons or hexagons, then exactly 12 of the faces must be pentagons.

[Hint: you may use Euler's polyhedron formula, which states that if a convex polyhedron has V vertices, E edges, and F faces, then $V - E + F = 2$.]

Question 2

The chemical elements phosphorus and chlorine can make two compounds: *phosphorus trichloride* (PCl_3) and *phosphorus pentachloride* (PCl_5).

A molecule of phosphorus trichloride has a trigonal pyramidal shape (an equilateral triangular-based pyramid). It has four vertices, with a phosphorus atom at one vertex and a chlorine atom at each of the other three vertices.

A molecule of phosphorus pentachloride has a trigonal bipyramidal shape (two equilateral triangular-based pyramids joined together by their equilateral bases). It has five vertices, with a chlorine atom at each vertex and a phosphorus atom at the centre of the equilateral triangle.

The diagram below shows a molecule of phosphorus trichloride (left) and a molecule of phosphorus pentachloride (right), with each phosphorus atom represented by a grey ball and each chlorine atom represented by a black ball.

Describe the symmetry operations of each molecule.

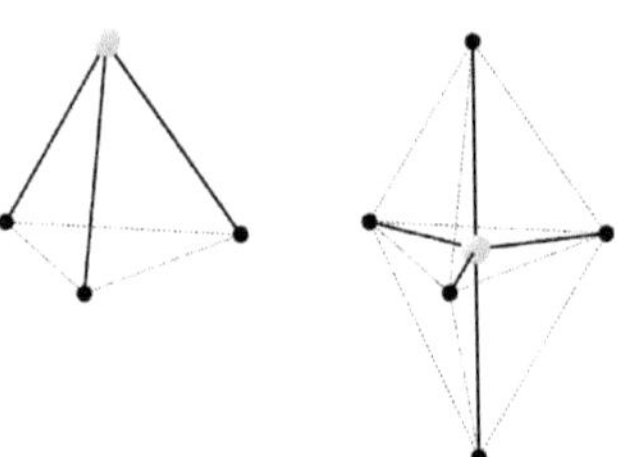

Question 3

Straight chain alkanes and *branch chain alkanes* are molecules which each consist of n carbon atoms and $2n + 2$ hydrogen atoms. Each carbon atom forms a single bond with each of exactly four other atoms, while each hydrogen atom forms a single bond with exactly one other atom. Show that in any such alkane, there cannot be a collection of bonds that form a circuit.

Group Round

Group Round

Senior Team Maths Challenge
2019 – 2020
National Final

Instructions

- Your team will have 40 minutes to answer 10 questions. Each team will have the same questions.
- Each question is worth 6 points. However, some questions are easier than others. Some questions are split into two parts worth 3 points each.
- You will have to decide your team's strategy for this group competition. Do you split up so that individuals work on a few questions each or do you work in pairs on a greater number of questions? Working all together on all the questions may well take too long.
- There is only one response sheet per team. Five minutes before the end of the time you will be told to finalise your answers and write them on the response sheet. Only this response sheet will be marked.

GROUP ROUND

SENIOR TEAM MATHS CHALLENGE
2019 – 2020
NATIONAL FINAL

QUESTION 1

What is the value of

$$\frac{5^8 - 3^8}{(5^4 + 3^4)(5^2 + 3^2)}?$$

[6 marks]

GROUP ROUND

SENIOR TEAM MATHS CHALLENGE
2019 – 2020
NATIONAL FINAL

QUESTION 2

a, b, c and d are numbers such that:

$$a : b = 2 : 7,$$
$$b : c = 2 : 11,$$
$$c : d = 2 : 13.$$

The value of d^2 is 1002001.

What is the value of a^2?

[6 marks]

Group Round

Senior Team Maths Challenge
2019 – 2020
National Final

Question 3

T, U and V are points on the circumference of a circle.

TUV is an equilateral triangle.

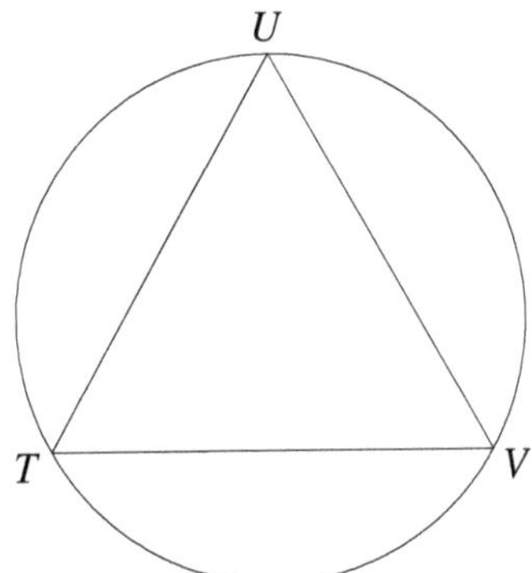

The circumference of the circle is 12 cm.

The area of the triangle TUV is $A\,\text{cm}^2$.

What is the value of $\pi^2 A$?

Give your answer in the form of $a\sqrt{p}$ where a is an integer and p is a prime number.

[6 marks]

Group Round

Senior Team Maths Challenge
2019 – 2020
National Final

Question 4

'3563' is an example of a number where every two consecutive digits form a multiple of 7.

The number N is a 500 digit number with the same property. The first digit of N is 8.

(a) What is the last digit of N? [3 marks]

There are M such 20 digit numbers with the property that every two consecutive digits form a multiple of 7 and the first digit is 8.

(b) What is the value of M? [3 marks]

GROUP ROUND

SENIOR TEAM MATHS CHALLENGE
2019 – 2020
NATIONAL FINAL

QUESTION 5

A sequence is defined as follows:

$$a_1 = 2,$$

and, for $n \geq 1$,

$$a_{n+1} = \text{the number of occurrences of } a_n \text{ in the first } n \text{ terms of the sequence.}$$

What is value of a_{1000}?

[6 marks]

Group Round

Senior Team Maths Challenge
2019 – 2020
National Final

Question 6

In this question the letters S, N, O, W, V, Y, E, T and I represent different digits.

The digits T, W, E, N, Y, V, I, S and O satisfy

$$\begin{array}{r} T\ W\ E\ N\ T\ Y \\ +\ T\ W\ E\ N\ T\ Y \\ \hline V\ I\ S\ I\ O\ N \end{array}$$

What is the value of $S + E + V + E + N + T + Y + T + W + O$?

[6 marks]

GROUP ROUND

SENIOR TEAM MATHS CHALLENGE
2019 – 2020
NATIONAL FINAL

QUESTION 7

Two numbers a and b satisfy the equations

$$a + b = 2,$$
$$a^2 + b^2 = 6.$$

(a) What is the value of $a^3 + b^3$? [3 marks]

(b) What is the value of $a^5 + b^5$? [3 marks]

GROUP ROUND

SENIOR TEAM MATHS CHALLENGE
2019 – 2020
NATIONAL FINAL

QUESTION 8

$ABCD, CPQR$ and $CWXY$ are squares.

R is the midpoint of AB.

The line QR intersects AD at Y.

The line PQ intersects XY at Z.

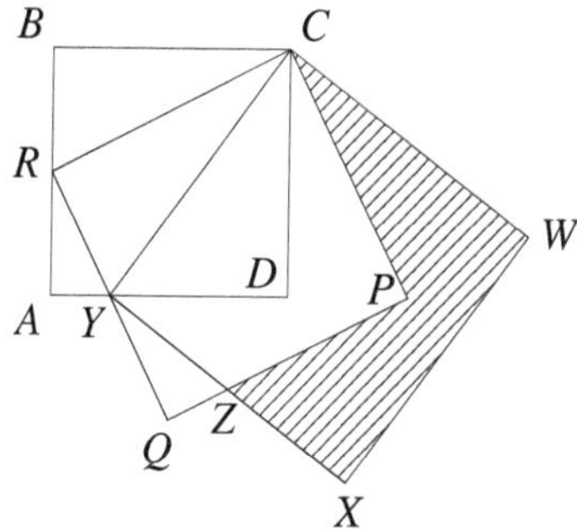

The area of the square $ABCD$ is 64.

What is the area of the shaded region $CWXZP$?

[6 marks]

GROUP ROUND

SENIOR TEAM MATHS CHALLENGE
2019 – 2020
NATIONAL FINAL

QUESTION 9

A *serrated* number is a positive number with the following property: for any three consecutive digits, the middle digit is either strictly greater than its two neighbours or strictly smaller than its two neighbours.

For example, 2020 is a serrated number, as is 31415, but 1243 and 314159 are not.

(a) How many 4-digit serrated numbers are there for which each of its digits is either 3, 2 or 1? [3 marks]

(b) How many 8-digit serrated numbers are there for which each of its digits is either 7, 6 or 5? [3 marks]

GROUP ROUND

SENIOR TEAM MATHS CHALLENGE
2019 – 2020
NATIONAL FINAL

QUESTION 10

Positive integers x, y and z satisfy the equations:

$$x^2 + y + z = 1988,$$
$$x + y^2 + z = 2020.$$

What is the value of z?

[6 marks]

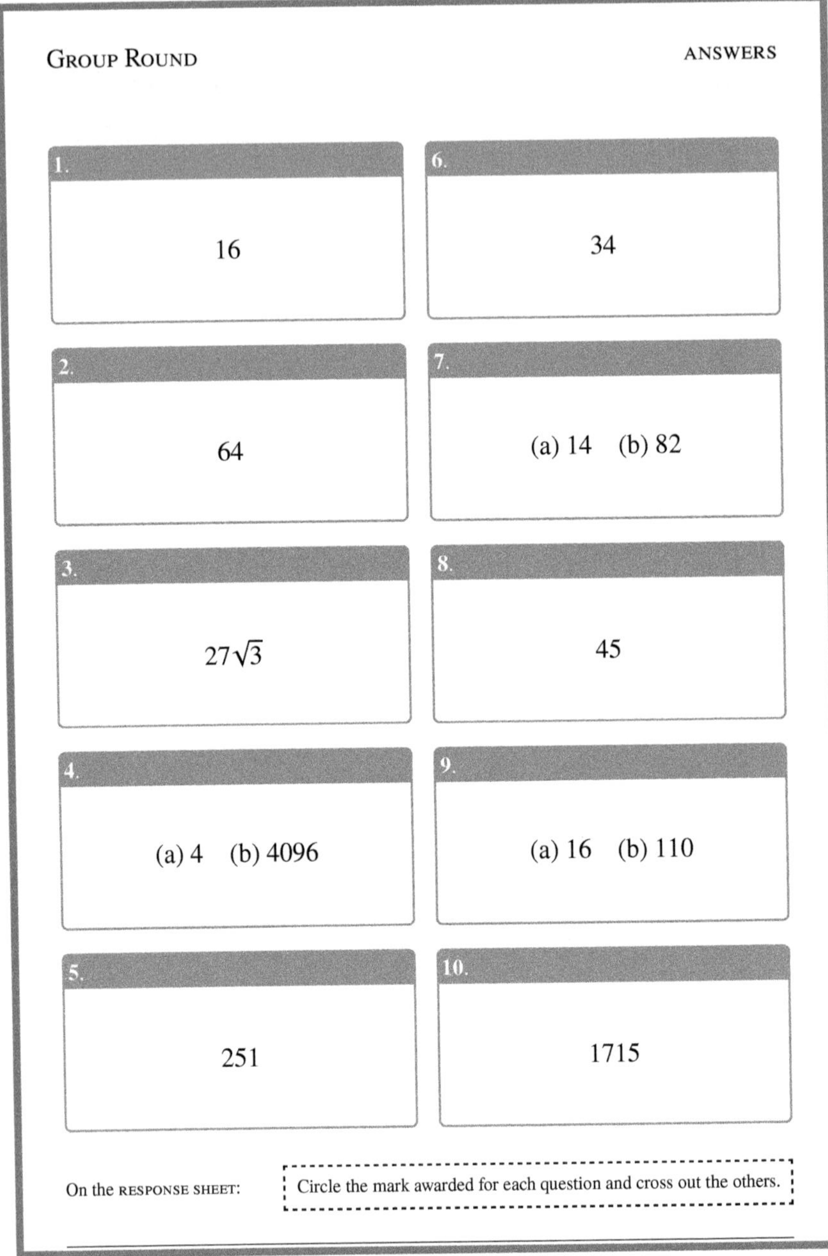

GROUP ROUND ANSWERS

	Answer		Answer
1.	16	6.	34
2.	64	7.	(a) 14 (b) 82
3.	$27\sqrt{3}$	8.	45
4.	(a) 4 (b) 4096	9.	(a) 16 (b) 110
5.	251	10.	1715

On the RESPONSE SHEET: Circle the mark awarded for each question and cross out the others.

Crossnumber

CROSSNUMBER

■	1		2	■	3		4		■
5	■	■	6			■		■	7
8		9		■	10			■	
	■		■	■		■	11	12	
13	14			■		■	■		■
■		■	■	15	■	16	17		18
19		20	■		■	■		■	
	■	21			■	22			
	■		■	23			■	■	
■	24				■	25			■

ACROSS

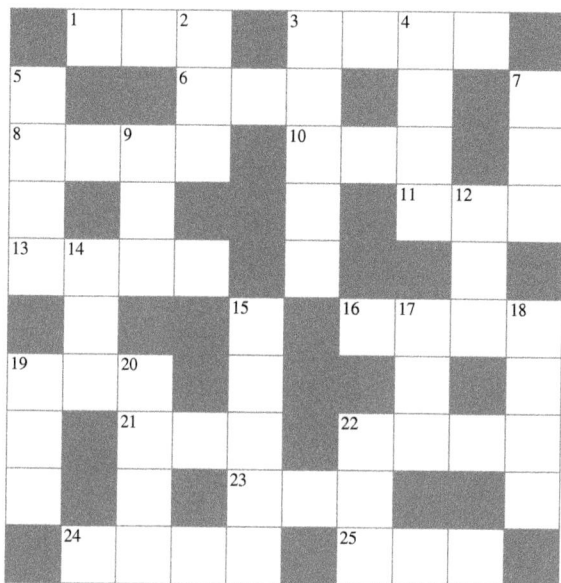

1. $x + 53$ where $5x + 6y = 2$ DOWN and $2x + y = 22$ DOWN (3)
3. A Fibonacci number (4)
6. The number of zeroes in $256^{100} \times 625^{200}$ (3)
8. The difference between 20 DOWN and 17 DOWN (4)
10. 6 ACROSS + 9 DOWN (3)
11. The number of different arrangements of the letters in the word NEWTON (3)
13. One tenth of the sum of the internal angles, in degrees, in a polygon with 22 DOWN sides (4)
16. A multiple of (100 more than 14 DOWN) (4)
19. $(15 + \sqrt{14})(15 - \sqrt{14})$ (3)
21. 125% of 23 ACROSS (3)
22. A multiple of a solution of $m^2 + 17m - 12$ DOWN $= 0$ (4)
23. 80% of 21 ACROSS (3)
24. The mean of 3 DOWN and 18 DOWN (4)
25. A factor of 18 DOWN (3)

DOWN

2. $(20 - \sqrt{20})(20 + \sqrt{20})$ (3)
3. A Fibonacci number (5)
4. A multiple of 19 ACROSS (4)
5. One twentieth of the sum of the internal angles, in degrees, in a polygon with 6 ACROSS sides (4)
7. The difference between 23 ACROSS and 25 ACROSS (3)
9. The hypotenuse of a triangle with shorter sides 108 and one-eighth of 11 Across (3)
12. 125% of 25 ACROSS (3)
14. 100 less than a factor of 16 ACROSS (3)
15. 3 ACROSS plus 3 DOWN (5)
17. 19 ACROSS is the mean of this number and 22 DOWN (3)
18. The number of different arrangements of the letters in the word HYPATIA (4)
19. An even multiple of the positive solution of $n^2 - 37n - 11$ ACROSS $= 0$ (3)
20. 3 ACROSS $- y$ where $19x + 4y = 19$ ACROSS and $7x + 6y = 1$ ACROSS (4)
22. The number of digits in $512^{16} \times 125^{48}$ (3)

Crossnumber

	[1] 1	2	[2] 3		[3] 1	5	[4] 9	7	
[5] 7			[6] 8	0	0		0		[7] 1
[8] 1	3	[9] 1	0		[10] 9	1	7		0
8		1			4		[11] 3	[12] 6	0
[13] 2	[14] 5	7	4		6			3	
	0			[15] 1		[16] 1	[17] 2	0	[18] 2
[19] 2	1	[20] 1		2			7		5
7		[21] 5	0	5		[22] 1	7	8	2
0		8		[23] 4	0	4			0
	[24] 6	7	3	3		[25] 5	0	4	

Shuttle

United Kingdom Mathematics Trust

Advanced Mathematics Support Programme

SENIOR TEAM MATHS CHALLENGE 2019–20

NATIONAL FINAL

SHUTTLE

A1

$$10! + 9! = 99 \times r!$$

Pass on the value of r.

Note: $m!$, (read 'm factorial'), where m is a positive integer, is the product of all the positive integers less than or equal to m. For example $5! = 5 \times 4 \times 3 \times 2 \times 1 = 120$.

United Kingdom Mathematics Trust

Advanced Mathematics Support Programme

SENIOR TEAM MATHS CHALLENGE 2019–20

NATIONAL FINAL

SHUTTLE

A3

T is the number you will receive.

$$8^n = 32^{(T-1)}.$$

Pass on the value of n.

United Kingdom Mathematics Trust

SENIOR TEAM MATHS CHALLENGE 2019–20

NATIONAL FINAL

SHUTTLE

A2

T is the number you will receive.

$a : b = 1 : 2, \quad b : c = 3 : 4, \quad c : d = 5 : 6, \quad d : a = x : (T+12).$

Pass on the value of x.

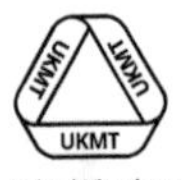

United Kingdom Mathematics Trust

SENIOR TEAM MATHS CHALLENGE 2019–20

NATIONAL FINAL

SHUTTLE

A4

T is the number you will receive.

The sequence $a_1, a_2, a_3, \ldots$ is defined by

$$a_1 = 1,$$
$$a_2 = T,$$
and, for $n > 1$,
$$a_{n+1} = \frac{a_n}{a_{n-1}}.$$

Write down the value of a_{2021}.

United Kingdom Mathematics Trust

Advanced Mathematics Support Programme

SENIOR TEAM MATHS CHALLENGE 2019–20

NATIONAL FINAL

SHUTTLE

B1

The sum of the first n even positive integers is 650.

Pass on the value of n.

United Kingdom Mathematics Trust

Advanced Mathematics Support Programme

SENIOR TEAM MATHS CHALLENGE 2019–20

NATIONAL FINAL

SHUTTLE

© UKMT 2019–20

B3

T is the number you will receive.

$$A = \left(\frac{1}{4}\right)^{\frac{T}{2}} \times \left(\frac{1}{8}\right)^{\frac{2}{3}} \times \left(\frac{1}{16}\right)^{-\frac{3}{4}} \times \left(\frac{1}{64}\right)^{-\frac{5}{6}}.$$

$$B = 27^{-\frac{2}{3}} \times 9^{\frac{T}{2}} \times 81^{-\frac{3}{4}}.$$

Pass on the value of $A + B$.

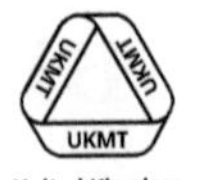

United Kingdom Mathematics Trust

SENIOR TEAM MATHS CHALLENGE 2019–20

NATIONAL FINAL

SHUTTLE

B2

T is the number you will receive.

A cylinder has volume $240\,\text{cm}^3$ and surface area $120\,\text{cm}^2$.

A similar cylinder has volume $1200T\,\text{cm}^3$ and surface area $24k^3\,\text{cm}^2$.

Pass on the value of k.

United Kingdom Mathematics Trust

SENIOR TEAM MATHS CHALLENGE 2019–20

NATIONAL FINAL

SHUTTLE

B4

T is the number you will receive.

The operation $\diamond$ is defined by

$$a \diamond b = ab + 3a + 2b.$$

Write down the value of

$$(T \diamond 1) \diamond 2.$$

United Kingdom Mathematics Trust

Advanced Mathematics Support Programme

SENIOR TEAM MATHS CHALLENGE 2019–20

NATIONAL FINAL

SHUTTLE

C1

m, $m + 2$, $m + 6$ and $m + 8$ are all prime numbers, with $m > 11$.

Pass on the least possible value of m.

United Kingdom Mathematics Trust

Advanced Mathematics Support Programme

SENIOR TEAM MATHS CHALLENGE 2019–20

NATIONAL FINAL

SHUTTLE

C3

T is the number you will receive.

Four vertices of a regular octagon are joined to form a square.

The length of each side of the octagon is $\sqrt{T}$.

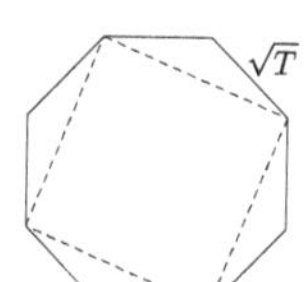

Find the area of the square in the form $a + \sqrt{b}$, where a and b are integers.

Pass on the value of $a + b$.

United Kingdom
Mathematics Trust

Senior Team Maths Challenge 2019–20

National Final

Shuttle

C2

T is the number you will receive.

Five integers are such that

the range is $T - 1$ more than the mean,
the mean is $T - 1$ more than the median, and
the median is $T - 1$ more than the mode.

$9(T - 1)$ is the largest of the five integers.

Pass on the mean of the other four integers.

United Kingdom
Mathematics Trust

Senior Team Maths Challenge 2019–20

National Final

Shuttle

C4

T is the number you will receive.

Let $x \diamond y = \sqrt{x^2 + 4y^2 - 4xy}$.

Write down the value of $(T \diamond 3) \diamond 5$.

United Kingdom Mathematics Trust

SENIOR TEAM MATHS CHALLENGE 2019–20

NATIONAL FINAL

SHUTTLE

D1

x, y and z are positive integers with $x < y < z$, and

$$x + 2xy + 3xyz = 71.$$

Pass on the value of z.

United Kingdom Mathematics Trust

SENIOR TEAM MATHS CHALLENGE 2019–20

NATIONAL FINAL

SHUTTLE

© UKMT 2019–20

D3

T is the number you will receive.

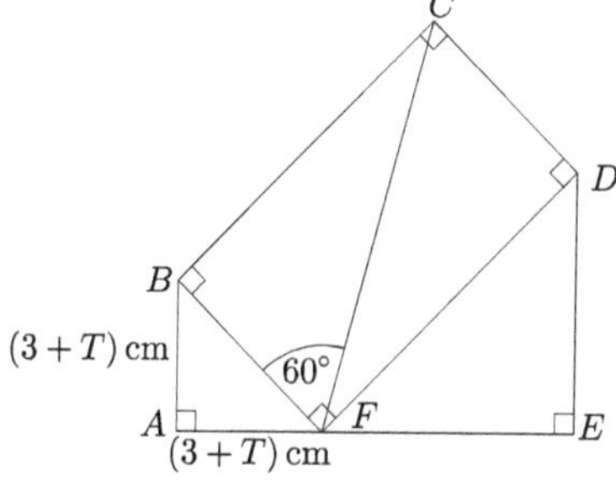

$BFDC$ is a rectangle and AFE is a straight line. Also $AF = AB = (3 + T)$ cm, $\angle BAF = \angle DEF = 90°$ and $\angle BFC = 60°$.

The length of DE is $\sqrt{k}$ cm. Pass on the value of $k + T$.

United Kingdom Mathematics Trust

Advanced Mathematics Support Programme

SENIOR TEAM MATHS CHALLENGE 2019–20

NATIONAL FINAL

SHUTTLE

D2

T is the number you will receive.

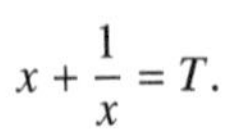

$$x + \frac{1}{x} = T.$$

$x^3 + \dfrac{1}{x^3}$ is the four-digit integer ‘$abcd$’.

Pass on the value of

$$(a + c) - (b + d).$$

Hint: $(p + q)^3 = p^3 + 3p^2q + 3pq^2 + q^3$.

United Kingdom Mathematics Trust

Advanced Mathematics Support Programme

SENIOR TEAM MATHS CHALLENGE 2019–20

NATIONAL FINAL

SHUTTLE

D4

T is the number you will receive.

The simultaneous equations

$$x^2 + Txy + \frac{T^2}{4}y^2 = 729$$

$$x^2 + \frac{T}{2}xy + \frac{T^2}{16}y^2 = 225$$

have four solution pairs (x, y).

Write down the sum of the two positive values of y.

SHUTTLE ANSWERS

A1	B1	C1	D1
8	25	101	11
A2	**B2**	**C2**	**D2**
64	5	400	0
A3	**B3**	**C3**	**D3**
105	3	320 800	27
A4	**B4**	**C4**	**D4**
$\frac{1}{105}$	74	320 784	8

On the RESPONSE SHEET: Circle the mark awarded for each question and cross out the others. At the end of the round, either circle the bonus mark or cross it out.

 SENIOR TEAM MATHS CHALLENGE 2019–20 NATIONAL FINAL

Relay

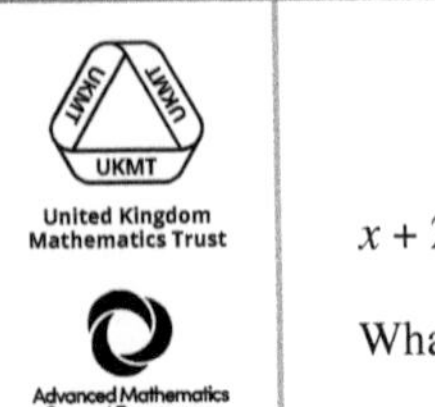

SENIOR TEAM MATHS CHALLENGE 2019–20

NATIONAL FINAL

RELAY

A1

$x + 2y + 3z = 4$ and $5y + 6z + 7x = 8$.

What is the value of $\dfrac{y}{x}$ for $x \neq 0$?

ANSWER:

United Kingdom Mathematics Trust

SENIOR TEAM MATHS CHALLENGE 2019–20

NATIONAL FINAL

RELAY

A2

Ten points are equally spaced around the circumference of a circle.

How many **different**, that is, non-congruent, triangles can be formed by joining three of these points?

ANSWER:

United Kingdom Mathematics Trust

SENIOR TEAM MATHS CHALLENGE 2019–20

NATIONAL FINAL

RELAY

A3

The operation $\diamond$ is defined by $x \diamond y = x^2 - 2y$.

What are the two solutions of the equation

$$(1 \diamond a) \diamond (a \diamond 7) = 125?$$

ANSWER:

United Kingdom Mathematics Trust

SENIOR TEAM MATHS CHALLENGE 2019–20

NATIONAL FINAL

RELAY

A4

The diagram shows a square $ABCD$ and a sector AEB of a circle with centre A and radius AB. DE meets AB at the point X.

AB has length 2 and $\angle BAE = \frac{1}{3} \times \angle DAB$.

Find the area, in square units, of the shaded region BXE.

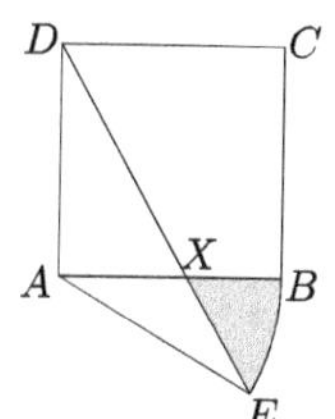

Express your answer in the form $a\pi + b\sqrt{3}$, where a and b are rational numbers expressed in their simplest form.

ANSWER:

United Kingdom Mathematics Trust

Advanced Mathematics Support Programme

SENIOR TEAM MATHS CHALLENGE 2019–20

NATIONAL FINAL

RELAY

A5

The positive integer k satisfies the equation

$$\sqrt{k} - \frac{12}{\sqrt{k}} = 3\sqrt{3}.$$

What is the value of k?

ANSWER:

United Kingdom Mathematics Trust

Advanced Mathematics Support Programme

SENIOR TEAM MATHS CHALLENGE 2019–20

NATIONAL FINAL

RELAY

A6

A sequence of positive integers, $a_1, a_2, a_3, \ldots$ satisfies $a_1 < a_2 < a_3 < \ldots$ and has the property that the difference between consecutive terms increases by 1 each time.

The sum of the first six terms is 158 and $a_7 = 22a_1$.

What is the value of the second term of the sequence?

ANSWER:

United Kingdom Mathematics Trust

Advanced Mathematics Support Programme

SENIOR TEAM MATHS CHALLENGE 2019–20

NATIONAL FINAL

RELAY

A7

'abc' is a three-digit integer, where a, b and c are three different digits.

$$\text{`}abc\text{'} = 23(a + b + c).$$

What is the integer 'abc'?

ANSWER:

United Kingdom Mathematics Trust

Advanced Mathematics Support Programme

SENIOR TEAM MATHS CHALLENGE 2019–20

NATIONAL FINAL

RELAY

A8

How many pairs of integers (x, y) are there such that

$$2x^2 - 2xy + y^2 = 25?$$

ANSWER:

United Kingdom Mathematics Trust

SENIOR TEAM MATHS CHALLENGE 2019–20

NATIONAL FINAL

RELAY

$ABCD$ is a square of side 2 with M the midpoint of AB.
A point P lies inside the square so that PM has length 1 and DP has length 2.

What is the distance of P from AD?
Give your answer either as a fraction in its simplest terms, or as a decimal.

ANSWER:

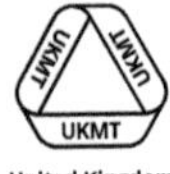

United Kingdom Mathematics Trust

SENIOR TEAM MATHS CHALLENGE 2019–20

NATIONAL FINAL

RELAY

Each cell of a 3×3 grid is to be filled with one of the digits 1, 2, 3 in such a way that each row and each column contains each digit exactly once.
The diagram below shows one such arrangement.

1	3	2
3	2	1
2	1	3

In how many different ways is it possible to fill the grid?

ANSWER:

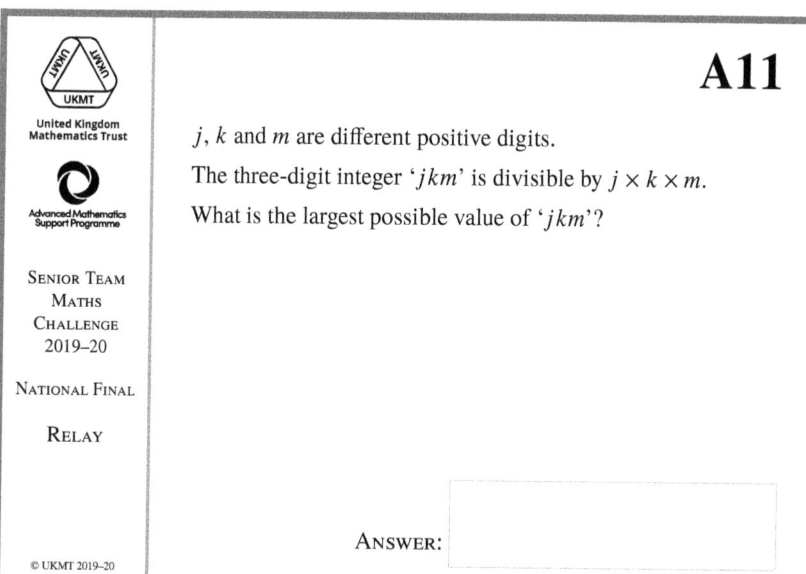

United Kingdom Mathematics Trust

Advanced Mathematics Support Programme

SENIOR TEAM MATHS CHALLENGE 2019–20

NATIONAL FINAL

RELAY

A11

j, k and m are different positive digits.

The three-digit integer 'jkm' is divisible by $j \times k \times m$.

What is the largest possible value of 'jkm'?

ANSWER:

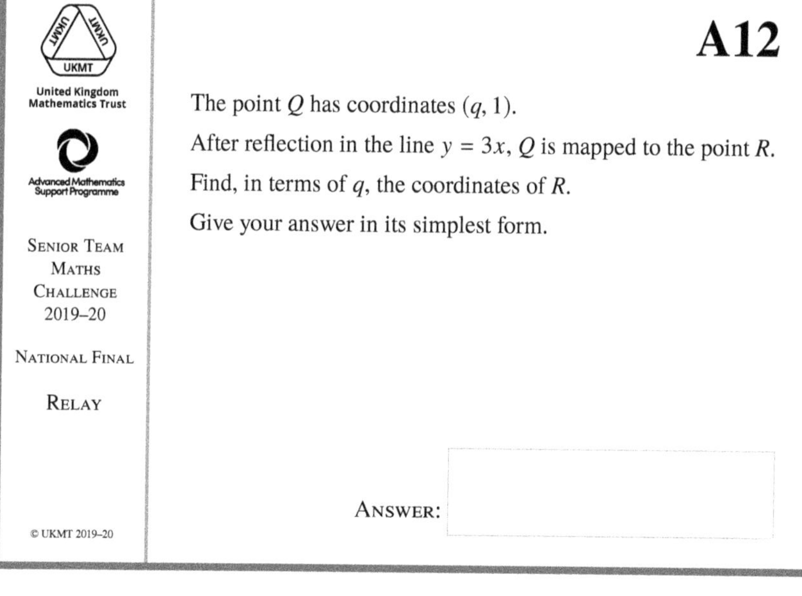

United Kingdom Mathematics Trust

Advanced Mathematics Support Programme

SENIOR TEAM MATHS CHALLENGE 2019–20

NATIONAL FINAL

RELAY

A12

The point Q has coordinates $(q, 1)$.

After reflection in the line $y = 3x$, Q is mapped to the point R.

Find, in terms of q, the coordinates of R.

Give your answer in its simplest form.

ANSWER:

United Kingdom Mathematics Trust

SENIOR TEAM MATHS CHALLENGE 2019–20

NATIONAL FINAL

RELAY

A triangle with sides in the ratio 7 : 24 : 25 is inscribed inside a circle of radius 5.

What is the area of the triangle?
Give your answer either as a fraction in its simplest terms, or as a decimal.

ANSWER:

United Kingdom Mathematics Trust

SENIOR TEAM MATHS CHALLENGE 2019–20

NATIONAL FINAL

RELAY

B2

$5x + 6y + 7z = 8$ and $9y + 10z + 11x = 12$.

What is the value of $\dfrac{z}{x}$ for $x \neq 0$?

ANSWER:

United Kingdom Mathematics Trust

Advanced Mathematics Support Programme

SENIOR TEAM MATHS CHALLENGE 2019–20

NATIONAL FINAL

RELAY

B3

The diagram shows an equilateral triangle ABC with sides of length 2, and a sector BAD of a circle with centre A and radius 2. $\angle CAB = 2 \times \angle BAD$.

The point M is the midpoint of AB. The line CM is extended to meet AD at the point E.

What is the area of the shaded region $BMED$?

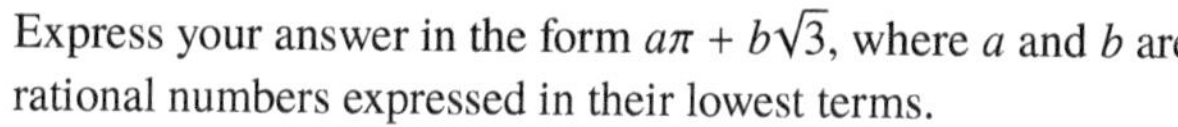

Express your answer in the form $a\pi + b\sqrt{3}$, where a and b are rational numbers expressed in their lowest terms.

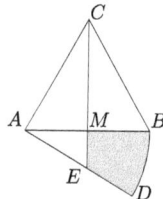

ANSWER:

United Kingdom Mathematics Trust

Advanced Mathematics Support Programme

SENIOR TEAM MATHS CHALLENGE 2019–20

NATIONAL FINAL

RELAY

B4

The operation $\diamond$ is defined by $x \diamond y = x - 3y$.

What are the two solutions of the equation

$$(a \diamond 4) \diamond (a^2 \diamond 4) = 0?$$

ANSWER:

United Kingdom Mathematics Trust

Senior Team Maths Challenge 2019–20

National Final

Relay

B5

'abc' is a three-digit integer, where a, b and c are three different digits.

$$\text{'}abc\text{'} = 17(a + b + c).$$

What is the integer 'abc'?

Answer:

United Kingdom Mathematics Trust

Senior Team Maths Challenge 2019–20

National Final

Relay

B6

The positive integer k satisfies the equation

$$\sqrt{k} - \frac{10}{\sqrt{k}} = \sqrt{5}.$$

What is the value of k?

Answer:

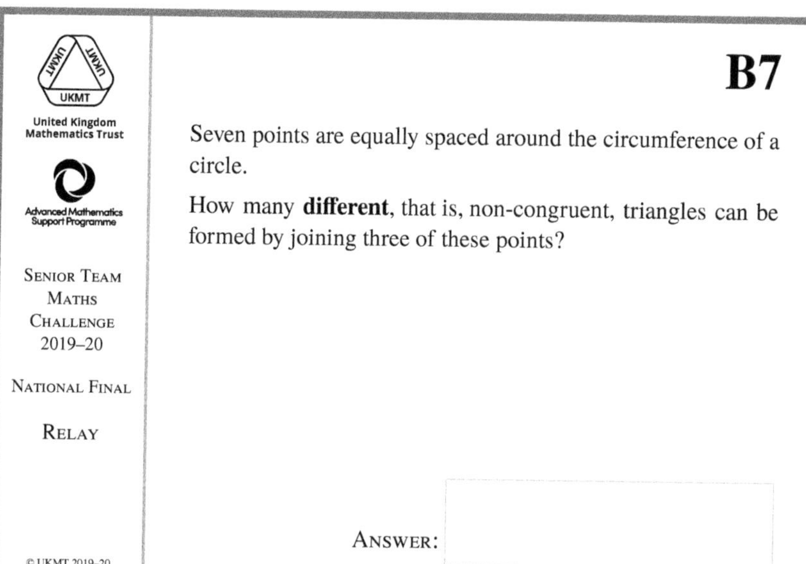

United Kingdom Mathematics Trust

Advanced Mathematics Support Programme

SENIOR TEAM MATHS CHALLENGE 2019–20

NATIONAL FINAL

RELAY

B7

Seven points are equally spaced around the circumference of a circle.

How many **different**, that is, non-congruent, triangles can be formed by joining three of these points?

ANSWER:

United Kingdom Mathematics Trust

Advanced Mathematics Support Programme

SENIOR TEAM MATHS CHALLENGE 2019–20

NATIONAL FINAL

RELAY

B8

How many pairs of integers (x, y) are there such that

$$x^2 - 4xy + 5y^2 = 169?$$

ANSWER:

United Kingdom Mathematics Trust

Advanced Mathematics Support Programme

SENIOR TEAM MATHS CHALLENGE 2019–20

NATIONAL FINAL

RELAY

B9

A sequence of positive integers, $a_1, a_2, a_3, \ldots$ satisfies $a_1 < a_2 < a_3 < \ldots$ and has the property that the difference between consecutive terms increases by 1 each time.

The sum of the first six terms is 137 and $a_8 = 9a_1$.

What is the value of the second term of the sequence?

ANSWER:

United Kingdom Mathematics Trust

Advanced Mathematics Support Programme

SENIOR TEAM MATHS CHALLENGE 2019–20

NATIONAL FINAL

RELAY

B10

Each cell of a 3×3 grid is to be filled with one of the digits 0, 1, 2 or 3 using the following conditions:

(i) each digit appears at least once within the grid,

and

(ii) each row contains exactly two zeros.

An example of such an arrangement is shown below:

0	2	0
0	0	1
0	0	3

In how many ways is it possible to complete the grid?

ANSWER:

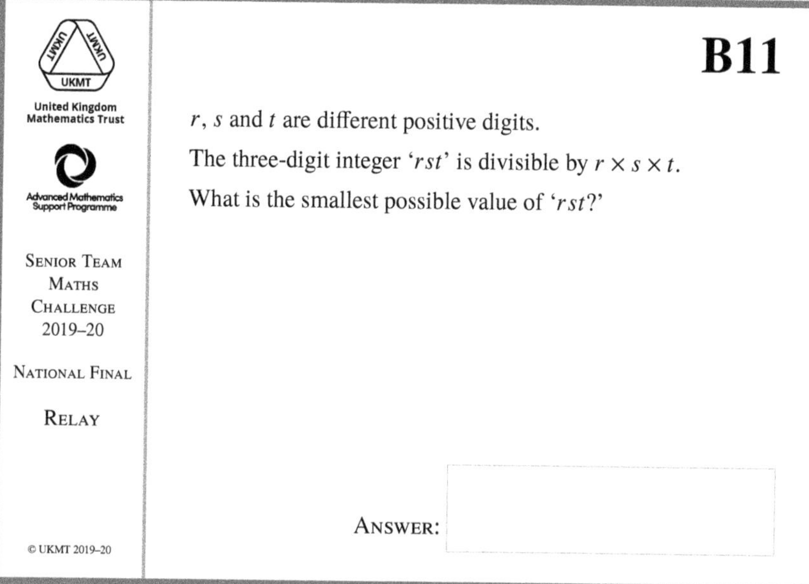
UKMT
United Kingdom Mathematics Trust
Advanced Mathematics Support Programme
SENIOR TEAM MATHS CHALLENGE 2019–20
NATIONAL FINAL
RELAY
© UKMT 2019–20
B11
r, s and t are different positive digits.
The three-digit integer 'rst' is divisible by $r \times s \times t$.
What is the smallest possible value of 'rst?'
ANSWER:

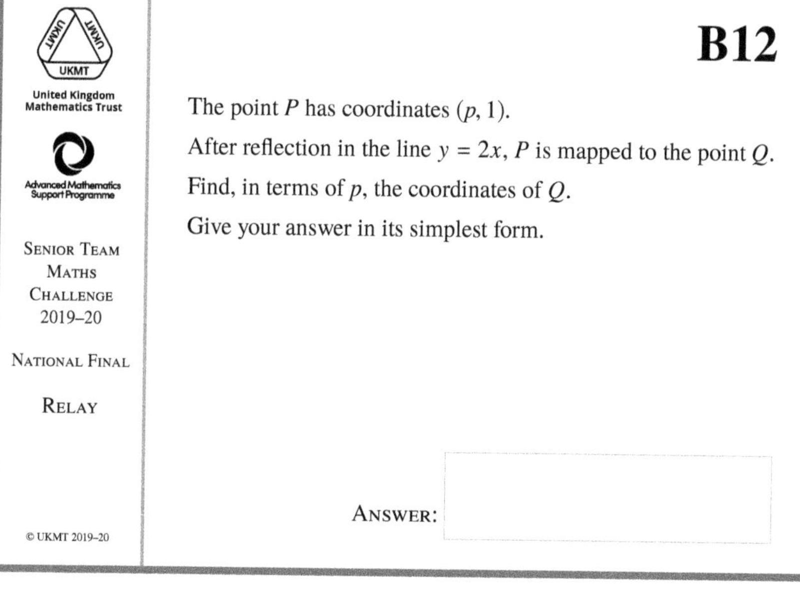
UKMT
United Kingdom Mathematics Trust
Advanced Mathematics Support Programme
SENIOR TEAM MATHS CHALLENGE 2019–20
NATIONAL FINAL
RELAY
© UKMT 2019–20
B12
The point P has coordinates $(p, 1)$.
After reflection in the line $y = 2x$, P is mapped to the point Q.
Find, in terms of p, the coordinates of Q.
Give your answer in its simplest form.
ANSWER:

RELAY SCORESHEET

TEAM NUMBER SCHOOL NAME

A1 -5 (0)(2)	**A5** 48 (0)(2)	**A9** $\frac{8}{5}$ or $1\frac{3}{5}$ or 1.6 (0)(2)
B1 $\frac{336}{25}$ or $13\frac{11}{25}$ or 13.44 (0)(2)	**B5** 153 (0)(2)	**B9** 12 (0)(2)
A2 8 (0)(2)	**A6** 11 (0)(2)	**A10** 12 (0)(2)
B2 7 (0)(2)	**B6** 20 (0)(2)	**B10** 162 (0)(2)
A3 -6 and 8 (0)(2)	**A7** 207 (0)(2)	**A11** 816 (0)(2)
B3 $\frac{1}{3}\pi - \frac{1}{6}\sqrt{3}$ (0)(2)	**B7** 4 (0)(2)	**B11** 128 (0)(2)
A4 $\frac{1}{3}\pi - \frac{1}{3}\sqrt{3}$ (0)(2)	**A8** 12 (0)(2)	**A12** $\left(\frac{3-4q}{5}, \frac{4+3q}{5}\right)$ or equiv (0)(2)
B4 $-\frac{8}{3}$ and 3 (0)(2)	**B8** 12 (0)(2)	**B12** $\left(\frac{4-3p}{5}, \frac{3+4p}{5}\right)$ or equiv (0)(2)

Correct answers score 2 points: circle 2 or 0 for each question and cross out the other number.
At the end of the round, draw a line under the last question attempted.

FINAL SCORE /48

 SENIOR TEAM MATHS CHALLENGE 2019–20 NATIONAL FINAL

Chapter 4

Enrichment activities

4.1 Mathematical Circles

Mathematical Circles enable enthusiastic mathematicians from a local area to come together for two days and follow a mathematically demanding programme. The Mathematical Circles are aimed at students in Year 10 (England and Wales), S3 (Scotland) and Year 11 (Northern Ireland). Mathematical Circles developed from two trial events in spring 2012. Following on from the success of these we have since significantly expanded the programme. Local schools are invited to select two students to participate in the two-day events, which consist of mathematically demanding work on topics such as geometry, proof and modular arithmetic. Students have the opportunity to discuss mathematics and make new friends from other schools in their area.
In 2019–20, Mathematical Circles only took place in Winchester (St Swithun's School) due to Covid-19.

4.2 National Mathematics Summer Schools

Our Summer Schools are designed for young people with an interest in mathematics. They are week-long residential events that aim to promote mathematical thinking and provide an opportunity for participants to meet other students and adults who enjoy mathematics.

At each summer school, 48 students take part in demanding and stimulating sessions led by volunteers. They are supported by a group of senior students who have previously attended a summer school. Time is set aside to allow students to work together informally on problems and to socialise.
Students in Years 10 or 11 (England and Wales), S3 or S4 (Scotland) and Years 11 or 12 (Northern Ireland) who are among the top 1.5 % of scorers in the Intermediate Mathematical Challenge are selected by lottery and invited to attend one of the summer schools. They must be aged 14–16 years at the time of the event. Only one student per school may be invited and only students resident in the UK are eligible. Schools are not invited to send students two years running to ensure, in a fair way, that students from as many different schools across the UK can take part in Summer Schools.
We were sadly unable to run in-person summer schools in 2020 due to Covid-19, so instead we ran a Virtual Summer School (VSS2020) to enable those students who would have otherwise missed out, to experience some of the elements of a UKMT summer school. The event took place from the 3-7 August 2020 and we ran three junior and two senior sessions each day, with a dedicated Google classroom space, virtual classes and lectures, enrichment exercises and a chat space. The event was led by Sue Cubbon and a team of dedicated summer school volunteers and feedback from students was extremely positive. 70 students took part.

4.3 Mentoring Scheme

Our Mentoring Scheme provides sets of challenging and engaging problems each month to help young people develop their problem solving skills. Students on the Mentoring Scheme work with a Mentor whose role is to provide encouragement and guidance about tackling the problems and writing solutions.
The Mentoring Scheme consists of nine programmes named as follows. They are intended to increase in difficulty and in the level of prerequisite knowledge required.

1. Pythagoras;
2. Hypatia;
3. Archimedes;

4. Mary Cartwright;
5. Julia Robinson;
6. Emmy Noether;
7. Hanna Neumann;
8. G. H. Hardy;
9. Srinivasa Ramanujan.

Each programme consists of a monthly set of problems running from October to May. It is intended that there is a progression in the difficulty and sophistication of problems from the beginning to the end within each programme.

The Olympiad Mentoring Scheme is run separately from the main Mentoring Scheme. and is for students who may take part in national and international competitions. Participation in the Advanced Olympiad Mentoring Scheme is by invitation.

4.4 Ri Mathematics Masterclasses Summer Celebration events

High-performing Intermediate and Senior Mathematical Challenge students in certain areas across the UK are usually invited to attend Summer Celebration events run jointly with the Royal Institution. In 2020, these events did not take place.

These events give students from Years 9 and 12 (England) and S3 (Scotland) the opportunity to attend inspiring lectures, meet mathematicians from their local area, and have a go at 'hands-on' mathematics.

Chapter 5

International competitions

5.1 Training and team selection

Various training camps are held throughout the year to select and prepare students for participation in the UK team at forthcoming International Mathematical Olympiads and other international events. In addition, the Advanced Mentoring Scheme supports team members and potential team members for particular international competitions.

5.1.1 Oxford Camp

The annual cycle starts with a week-long camp in September for students who are new to Olympiad training, and who may be ready to join the Olympiad Mentoring Schemes. Students for this camp are selected based on information from various competitions and activities held over the previous year, including those for younger students.

5.1.2 Hungary Camp

Since 2001, there has been a visit to Hungary over the New Year for a joint camp. Students for this camp are selected based on performances in the British Mathematical Olympiad Round 1, taking into account how much previous training they have received.

5.1.3 Trinity Training Camp

The traditional Easter training and selection camp has been held at Trinity College, Cambridge from 1989 onwards. Based on performances in the British Mathematical Olympiad Rounds 1 and 2, along with any other relevant information such as that from exams at the Hungary Camp, 20 or so students are invited to the Trinity Camp. There are four or five sessions each day covering material which is useful for IMO problems. Part way through and at the end of the five-day camp, students sit two 4.5-hour IMO-style papers, after which a final small UK IMO squad is selected for intensive training.

5.1.4 Pre-IMO Camp

A final training camp for the IMO team is held immediately before the IMO. This is usually a joint camp with the Australian team, where the Mathematics Ashes competition is held between the two countries.

5.2 Romanian Master of Mathematics

The UK has competed in the Romanian Master of Mathematics (RMM) since 2008 and sends six contestants to this competition, which is normally held in late February. Further details about the RMM, including past papers, can be found at bmos.ukmt.org.uk/home/rmm.shtml.
In 2020, the Romanian Master of Mathematics was held on 25 February - 2 March in Bucharest.
The members of the UK team were:

Wilfred Ashworth	Sutton Grammar School for Boys
Sarah Gleghorn	Skipton Girls' High School
Yuhka Machino	Millfield School
Aron Thomas	Dame Alice Owen's School
Tommy Walker Mackay	Stretford Grammar School
Sherman Yip	Tonbridge School

This year the team came 5th out of 19 competing teams. They collected one Gold medal, two bronze medals and three Honourable Mentions. More information can be found at www.imo-register.org.uk.

5.3 European Girls' Mathematical Olympiad

The first European Girls' Mathematical Olympiad (EGMO), organised jointly with Murray Edwards College, Cambridge, was held in April 2012. This is now an annual international competition held in April. The UK team of four people is selected on the basis of results in the British Mathematical Olympiad Round 1 and 2.
The Mathematical Olympiad for Girls helps to identify students to engage in training for EGMO. Further details, including past papers, can be found at bmos.ukmt.org.uk/home/egmo.shtml.
The 9th European Girls' Mathematical Olympiad was held remotely on 15-21 April 2020 due to Covid-19. The members of the UK team were:

Naomi Bazlov	King Edward VI High School for Girls, Birmingham
Sarah Gleghorn	Skipton Girls' High School
Yihua Luo	St George's School for Girls
Yuhka Machino	Millfield School

The team came home with one gold, three silver medals, and were 5th out of 52 participating teams.

5.4 Balkan Mathematical Olympiad

The UK has competed as a guest team in the Balkan Mathematical Olympiad since 2005. No UK team member competes more than once in the Balkan Mathematical Olympiad; when sending teams to this competition, we aim to extend opportunities to participate in international competitions to as many students as possible. Further details, including how to view past papers, can be found at `bmos.ukmt.org.uk/home/balkan.shtml`.
In 2020, the Balkan Mathematical Olympiad was postponed to October 31st to 5th November 2020, due to Covid-19.

5.5 International Mathematical Olympiad

The International Mathematical Olympiad (IMO) is the pre-eminent mathematics competition for secondary-school students and is held

annually in a different country. The first IMO was held in 1959 in Romania, with 7 countries participating. It has gradually expanded to over 100 countries from six continents. For more information see www.imo-official.org.
The 61st International Mathematical Olympiad was hosted from Saint-Petersburg, Russia and held as a virtual competition from venues around the world on 21 and 22 September 2020. The members of the UK team were:

Samuel Liew	The West Bridgford School
Yuhka Machino	Millfield School
Benedict Randall Shaw	Westminster School
Aron Thomas	Dame Alice Owen's School
Tommy Walker Mackay	Stretford Grammar School
Sherman Yip	Tonbridge School

The first reserve for the IMO 2020 was Thomas Frith.
This year the team achieved one Gold, four Silver and one Bronze medal. The team finished 9th out of 105 participating teams. Yuhka Machino, the UK's Gold medallist, came 1st amongst all female participants and is the only female who was awarded a Gold medal in the IMO 2020.

Chapter 6

Volunteers

6.1 Governance

6.1.1 Trustees

A F Baker
Dr D Crawford[*]
R M Dorris[†]
R J Gazet
C James
G Keniston-Cooper
M C D Knapton
Dr G Leversha
J Ramsden MBE
Prof. A Rucklidge
Dr G C Smith

6.1.2 Office bearers

Chair	G Keniston-Cooper
Treasurer	Dr D Crawford
Secretary	M T Fyfe
Vice-Chairs	R J Gazet and Dr G C Smith

[*] Representative of the Mathematical Association (a Participating Body). [†] Representative of The Royal Institution of Great Britain (the Parent Body).

6.1.3 Members of the Trust

As of 31 July 2020, the Members of the Trust were the Trustees (listed above) together with:

R M Atkins
P Beckett
Dr R W Bray
Prof. J Brindley
Prof. C Budd OBE
D Bunnell
Dr C M Campbell
Dr K M Chicot
Dr D J Collins
Dr J D Cranch
A Crews
S Cubbon
M Dennis
Dr A Eames-Jones
Dr C C Fiddes
K Fogden
M T Fyfe
Dr A D Gardiner
H Groves MBE
K Hayward-Bradley
T J Heard MBE
F Heywood
P Hunt
Dr A Inglis
S Jameson-Petvin
Dr V Kadelburg
Dr V Kathotia
Dr C Kilgour
Dr E Kimber
P King
Prof. F C Kirwan
Prof. T W Körner
Prof. I B Leader
N Lord
Dr A J S Mann
P Matheson
Prof. A C McBride
S J Mulligan
P Murray
Dr J S Myers
Dr V Neale
Dr P M Neumann
D J Orton
L Piper
S Power
Dr A Randolph
P Ransom MBE
H Reeve
Prof. J C Robson
D Rowland
Dr A Sanders
Dr S Santos
Sir Prof. B Silverman
Dr J R Silvester
Dr A B Slomson
R C Smart
Dr C G Smith MBE
W B Stewart
P A J Thomas
A L Voice
J R Welham
R Wiltshire
M Wimbury
Dr D Yeo

6.1.4 Audit and Finance Committee

M Knapton (Chair)
Dr D Crawford
J Ramsden MBE

6.1.5 Investment Committee

G Keniston-Cooper
Dr D Crawford
P Jeffreys
Prof. A McBride (Chair)
S Yun-Farmbrough (from May 2020)

6.1.6 Nominations and Remuneration Committee

Dr K Chicot
C James
Prof. F Kirwan
Dr V Neale
Prof. A Rucklidge (Chair)

6.1.7 Risk and Compliance Committee

Prof. C Budd
R Dorris
J Gazet
S Hukovic (Chair)

6.2 Executive Committees

6.2.1 British Mathematical Olympiad

Dr J Cranch
Dr C Fiddes
J Gazet
Dr V Kadelburg (Secretary)
Prof. I Leader
Dr J Myers
D Rowland
Dr G Smith MBE (Chair)
Dr D Yeo

6.2.2 Challenges

A Baker
Dr D Crawford
K Fogden
H Groves MBE
C James (Chair)
Dr C Kilgour (Treasurer)
Prof. A McBride
P Murray
J Ramsden MBE (Secretary)
Dr A Randolph
Prof. C Robson
Dr A Slomson
R Wiltshire

6.2.3 Team Maths Challenges

P Beckett
Dr R Cretney (Chair from Nov 19)
M Dennis
P Hunt
Dr A Inglis
P Matheson (Secretary)
S Mulligan (Chair until Nov 19)
H Reeve

6.2.4 Enrichment

Dr K Chicot
S Cubbon (Vice Chair)
Dr L Kimber (Secretary)
Dr V Neale (Chair)
L Piper
Dr A Slomson

6.2.5 Publications

R Bhattacharyya
M.T Fyfe
J Gazet
Dr C Kilgour
Dr G Leversha (Chair)
N Lord
S Power
D Rowland

6.2.6 British Mathematical Olympiad Committee

R Atkins
R Bhattacharyya
P Coggins
Dr D Collins
M.T Fyfe
Prof. B Green
Dr J King
P King
Dr G Leversha
Prof. A McBride
Dr D Monk
Dr V Neale
Dr P Neumann
Dr P Pears
Dr A Sanders
Dr Z Stoyanov
Dr A West

6.3 Problem setters

6.3.1 Solo competitions

Junior and Intermediate Mathematical Challenges

D Crawford (Vice Chair)
K Fogden
C Gainlall
T Gardiner
H Groves MBE (Chair)
D Hamilton
P Murray
S Power
P Ransom MBE
A Slomson
A Voice

Senior Mathematical Challenge

S Barge
P Beckett
D Bunnell
K Fogden (Chair)
H Griffiths
H Groves MBE (Vice Chair)
D Haynes
S Maltby
C Oakley
F Shen
A Slomson

SMC used some of the late Andrew Jobbings' questions that he had composed for the 2020 paper.

Cayley, Hamilton and Maclaurin Olympiads

T Bowler (Vice-Chair)
D Bunnell
J Cranch
A Gardiner
M Griffiths
D Griller (Chair)
H Groves MBE
J Hall
G Leversha
A Parkinson
S Power
C Ramsay
F Shen
M Walker
D Winn

British Mathematical Olympiad

E Backhausz
E Beatty
L Betts
T Bowler
D Griller
S Haring
D Rowland (Chair)
G Smith MBE

Mathematical Olympiad for Girls

T Bowler
E Pound
V Kadelburg

6.3.2 Team Maths Challenge

Starters	A Parkinson
Group	D Bunnell
Crossnumber	P Ransom MBE
Shuttle	D Crawford, K Hayward-Bradley
Relay	A Ault, P Colville, S Essex
Poster	A Ault, C Campbell, P Beckett, R Cretney, T Gill, F Heywood, Dr M Walker
Checking	F Chalmers, W Dersley, A Inglis (Curriculum Consultant, Scotland), M Lawley (Lead Checker), S Mulligan (Chair, Team Maths Challenges Subtrust), J Ramsden MBE, S Saddique (Curriculum Consultant, Wales), A Slomson, E Toman (Curriculum Consultant, Northern Ireland)
Typesetting	F Heywood, S O'Hagan

6.3.3 Senior Team Maths Challenge

Group	C Oakley (Round Ruler), D Bunnell, K Fogden, D Hamilton, K Brown
Crossnumber	P Ransom MBE (Round Ruler)
Shuttle	P Beckett (Round Ruler), Dr D Crawford, P Healey, A Ginty
Relay	D Bunnell (Round Ruler), M T Fyfe (Round Ruler), P Walter, M Griffiths
Poster	R Cretney (Round Ruler) A Hewitt (AMSP), M Walker, C Campbell, P Neumann, T Rome (AMSP)
Checking	A Slomson (Lead Checker) M Dennis (Lead Checker), H Reeve, J Ramsden MBE, P Richmond, S Saddique

6.3.4 Primary Team Maths Resources

D Ball
A Bell
E Bull
D Bunnell
S Essex
R Greenhalgh
F Heywood
T Lunel
D Pinshon (Chair)
L Piper
A Slomson

6.3.5 Mentoring Committee

Dr A Andrews (Chair)
Dr J Gilbey
D Phillips
J Slater
Dr D Yeo

6.4 Markers, data inputters and judges

6.4.1 Cayley, Hamilton and Maclaurian Olympiad

R Bhattacharyya
T Bowler (Intermediate Olympiad Marking Coordinator)
P Coggins
J Cranch
A Gardiner
G Gardiner
D Griller
J Hall
S Haring
C James
V Kadelburg
G Leversha
S Power
D Rowland

6.4.2 British Mathematical Olympiad Round 1

O Hidalgo Arevalo
E Backhausz
T Backhausz
C Bambridge-Sutton
A Banerjee
S Bealing
E Beatty
J Beckett
P Beckett
J Bell
L Betts
R Bhattacharyya
T Bowler
A Carlotti
P Coggins
A Conmy
J Cranch
A Darby
C Eagle
R Flatley
R Freeland
A Gardiner

J Gazet
D Griller
A Gunning
S Haring
T Hennock
L Hill
I Hughes
I Jackson
S Jayasekera
V Kadelburg
H Khan
P King
W Li
H Metrebian
J Myers
T Pelling
E Pound
M Quail
Z Randelovic
D Rowland
A Shah
G Smith MBE
J Watson
P Winter
H Yau
R Zhou

6.4.3 British Mathematical Olympiad Round 2

E Beatty
J Bell
A Carlotti
A Chlebikova
J Cranch (Lead Marker)
R Freeland
J Gazet
J Haslegrave
F Illingworth
S Law
J Long (graphics)
G Majury
D Mestel
J Millar
J Myers
J Owladi
P Parmar
E Pound
D Rowland
A Sarkovic
Z Stoyanov
K Warburton
H Yau
D Yeo

6.4.4 Mathematical Olympiad for Girls

A Aggarwal
O Hidalgo Arevalo
A Asad
L Beatson
E Beatty
P Beckett
R Bhattacharyya
A Carlotti
P Coggins
J Cranch
MT Fyfe
V Kadelburg
J King
J Myers
M Orr
E Pound
M Quail
Z Randjelovic
A Rout
A Šarković
A Shah
A Stavrou

S Tate
K Warburton
J Watson

6.5 Inputters

C Beckett
N Turner

6.6 Mentors

L Franz
J Elmes
B Vlaar
O Hidalgo Arevalo
P Ransom MBE
A Thorley
R Atkins
S Hewitt
A Slomson
E Doman
P Price
L Samuelson
T Hillman
W Li
R Carling
S Biswas
P Scott
S Tate
K Klein
C Schafer
L Agarwal
P Czerniawski
L Hollom
D Berry
A Salagean
P Healey
H Strauss
P Beckett
J Hall
P Voutier
G Craciun
F Feser
C Ellingham
M Illing
M Corte-real Santos
A Konstantinou
M Lipton
R Kilby
J Fawkes
N Behague
H Pradip
B Ford
C Marriott
M Penn
B Cameron
Y Wang
D Hamilton
P Lloyd
I Hughes
M Knapton
J Gilbey
H Yu Wu
Z Dennison
M Cowley
D Low
K Miller
B Walker
M Eyre-Morgan
R Mendoza Smith
B Spells
J Ripp
T Baycroft
D Yeo
A Sanders
A Banerjee
E Beatty
F Illingworth
H Yau
J Aaronson
J Benton
J Myers
K Warburton
N Nanda
R Zhou
R Freeland
R Cates
T Read
G Leversha
E Backhausz
M Quail
J Gazet
S Bealing
S Kittle
MT Fyfe
B Elkins
J Cullen
R Talbut
B Elkins
C Lewis-Brown
O Thomas
G McCaughan
M Wildon
C Athorne

P Withers
J Bamford
C Luke
C Shafer
P Ramsey
K Porteous
J Lazic

6.7 Event leaders

6.7.1 Team and Senior Team Maths Challenge Regional Coordinators and Helpers

H Ainsley
P Andrews
A Ault
M Bailey
A Baker
G Baker
B Ballantyne
P Beckett
A Bell
S Biswas
Z Booth
N Bray
M Burrows
E Bull
K Burnham
R Cave
F Chalmers
R Chapman
P Colville
J Cranch
D Crawford
R Cretney
L Daniels
M Dennis
W Dersley
B Elkins
G Engelhardt
S Essex
S Fernandes
J Fox
L Franz
R Fraser
M T Fyfe
C Gainlall
H Gauld
T Gill
P Healey
T Heard
F Heywood
S Hughes
S Anne Huk
P Hunt
M Illing
A Inglis
K Klein
T Lunel
P Matheson
L McIntosh
M Miller
S Mulligan
H Mumby
P Noble
D Pinshon
V Pinto
V Pinto
L Piper
P Price
H Reeve
V Ridgman
S Saddique
J Slater
A Slomson
A Strong
L Swarbrick
P Thompson
E Toman
N Turner
L Walton
J Welham
T Whalley
I Wiltshire
R Wiltshire

6.7.2 National Mathematics Summer Schools

The 52nd National Mathematics Summer School

J Cranch
R Freeland
M T Fyfe
J Hodkinson
E Hull
D Phillips (Pastoral deputy)
C Ramsay (Director)
A Sharma
R Shiatis
A Slomson
B Vlaar
H Yorston

The 53rd National Mathematics Summer School

R Bhattacharyya (Director)
M Bradley
A Chlebikova
G Leversha
C Luke
G Majury (Pastoral deputy)
E Pound
P Russell
R Shiatis
L Wells

The 54th National Mathematics Summer School

C Bambridge-Sutton
A Baker (Pastoral deputy)
P Beckett
S Cubbon (Director)
J Cranch
D Crawford
S Jayasekera
L Piper
A Slomson
S Tate
M Weiserova

6.7.3 Training camps and international competitions

Oxford Camp

J Gazet (Director)
L Cullen
M Davies
O Feng
F Illingworth
G Majury
P Russell
K Warburton
D Yeo

Romanian Master of Mathematics

F Illingworth (Leader)
J Gazet (Deputy Leader)
G Majory (Pastoral)

Hungary Camp

E Beatty
D Yeo
K Warburton

Trinity Training Camp

J Benton
R Bhattacharyya
A Caraceni
J Gazet
G Gendler
F Illingworth
V Kadelburg
I Leader
N Nanda
P Russell
K Warburton
D Yeo (Virtual Camp Leader)

European Girls' Mathematical Olympiad

J Owladi (Leader)
K Warburton (Deputy Leader)

Balkan Mathematical Olympiad

R Bhattacharyya (Leader)
K Warburton (Deputy Leader)

International Mathematical Olympiad

F Illingworth (Academic Observer)
K Warburton (Deputy Leader)
D Yeo (Leader)

Chapter 7

Our supporters

We gratefully acknowledge the support of our major supporters.

Thanks go to our other supporter:
University of Bath

Sponsorship
XTX Markets
LetterOne

Donations
Dulwich College
The Haberdashers' Aske's Boys' School
Merton College, University of Oxford
St John's College, University of Cambridge

[XTX]
MARKETS

Lightning Source UK Ltd.
Milton Keynes UK
UKHW020639141221
395640UK00012B/820